Ruth Elwin Harris began storytelling when, during the last war, she and her brother went to stay with their grandfather in his isolated Somerset house: "We led a very solitary existence," she says. "Not that we minded. We read a lot and made up stories to entertain each other. We both loved the house and were very happy."

It is that house, christened Hillcrest, which provides the setting for *The Silent Shore,* the first book in the *Quantocks Quartet.* "My grandfather bought it in the 1930s from three elderly sisters — all of whom had been painters. Their murals still remained on the stable walls. I used to think about those sisters and wonder about life in the village when they were young. I started thinking what it would have been like if there had been another sister who didn't paint." Gradually that fourth sister turned into Sarah, the youngest of the Purcell girls and the heroine of *The Silent Shore.*

Ruth Elwin Harris lived abroad for several years before and after her marriage — at one time accompanying her husband to India where she helped out at an orphanage run by Mother Teresa's nuns. She now lives with her family in North Yorkshire. She has written stories for magazines and radio and three novels in the *Quantocks Quartet: The Silent Shore* (shortlisted for The Observer Teenage Fiction Award), *The Beckoning Hills* (published by Walker in paperback 1989) and *The Dividing Sea* (Julia MacRae Books 1989).

Also by Ruth Elwin Harris

The Beckoning Hills

The Silent Shore

Ruth Elwin Harris

WALKER BOOKS
LONDON

First published 1986 by Julia MacRae Books
This edition published 1989 by Walker Books Ltd
87 Vauxhall Walk, London SE11 5HJ

© 1986 Ruth Elwin Harris
Cover illustration by Emma Chichester-Clark

Printed in Great Britain by Cox and Wyman Ltd, Reading

British Library Cataloguing in Publication Data
Harris, Ruth Elwin
The silent shore.
I. Title
823'.914[F]   PR6058.A691
ISBN 0-7445-1313-8

"And, when the stream
Which overflowed the soul was passed away,
A consciousness remained that it had left,
Deposited upon the silent shore
Of memory, images and precious thoughts,
That shall not die, and cannot be destroyed."

from Wordsworth, *The Excursion*, Book 7

I would like to thank the many people who so willingly gave their time and shared their memories with me, singling out for especial thanks Loveday Cogan, St. John Couch, Hilary Dunn, D. N. Glenday, and Annie Hancock.

Contents

1910

Chapter One

They lurked in the shadows, all the evil spirits of every fairy story that she had ever read, ghouls, goblins, imps, crowding round the edges of the room, waiting for her to move into the darkness. One step outside the safe circle of lamplight and she would be trapped, caught in shadowy arms, carried away.

Even in her most terrible dreams she knew that in some way, at some time, there would be an awakening. Not tonight. Tonight was worse than any nightmare. She could not pretend tonight that her body was lying in bed upstairs. She knew it was here, in the kitchen, shivering on a stool, while Annie, like a witch at her cauldron, took the brightly coloured garments from a pile on the table and dropped them one by one into the bubbling copper on the range. One by one, the happy colours became the dreary black of night until only one garment remained.

"Not my yellow dress!" She fell off the stool, grabbed Annie's arm. "Please, Annie, not that. I don't want black. I hate black. Please, please, let me keep my yellow dress."

"Oh, Miss Sarah," Annie said, "you can't wear yellow – not at a time like this. 'Twouldn't be respectful. Let me have it now, there's a good girl."

"I won't, I won't!" But her voice wavered as she struggled

to hold the material. It was no good, she should have known it would be no good. Her lips trembled as the dress disappeared into the dye. Now there was nothing left; even her favourite dress had gone. She put her head down on the soft, smooth surface of the table and wept.

Annie pulled up a stool. "It's only for a little while, wearing black," she said gently. "You can have all the bright colours you want after."

"It was forever when Father died."

"It only seemed that way." She stroked the child's hair. "It's a mark of respect, you see, for your mother. And to let people know you're sad."

Sarah lifted her head. "She wanted to die. The doctor said she'd get better but she wanted to die. I heard you tell him so. It's her fault she's dead."

"Oh, love. You shouldn't listen to things like that. I didn't mean . . ." Annie searched for words. "She . . . she lost heart, you see, when your father died. She wanted to be with him."

"But Father's all right. Father's got God." Tears trickled down into Sarah's mouth, warm, salty. "Why didn't she think of us? We haven't got anybody. What's going to happen to us?"

"I don't know. No-one tells Annie things like that." She tried to smile. "Don't you worry, love. Mr. Mackenzie'll make sure you're all right. And your sisters are coming home tomorrow. You'll feel better then. Now, how about a nice cup of cocoa before I take you up to bed?"

The mug warmed Sarah's hands. She was safe on Annie's lap, with Annie's arms around her. She sipped the cocoa slowly, delaying the moment when she would have to pass that door on her way to bed.

"Not Miss Sarah," Annie had said. "There's no call for it. She's fanciful at the best of times – gets funny ideas in her head. We'll have bad nights for weeks. Let the child remember her mother as she was."

But Mrs. Mackenzie knew best. Mrs. Mackenzie always knew best. Grasping Sarah's hands, so tightly that the rings twisted into Sarah's fingers and made dents in her flesh, Mrs. Mackenzie had taken her through the door into the dining-room.

Don't look, Sarah had told herself, fixing her gaze on the rich maroons and peacock blues of the carpet that Father had brought back from Constantinople. First Father had gone, now Mother. But Mother had wanted to die, had wanted to be with Father. Perhaps she had even asked God to make her die. Sarah thought of her own prayers, of the shivering hours spent on her knees by the side of her bed, the promises to be good, for ever and ever, if only God would help. God hadn't listened to her. He had listened to Mother instead.

Sarah looked at her cocoa now, and wondered. How did God decide whose prayers to answer?

"Come on, love," Annie said. "It'll go cold if you don't drink it up quick."

Skin had formed on the surface, purply-pink and wrinkled, soft against her lips. When she drank, it separated into little pieces that caught between her teeth and on her tongue.

"Carry me, Annie."

"A big girl, like you! Seven – you're too heavy for poor Annie to carry."

Annie's arms held her secure. She rubbed her face against Annie's cheek, still damp from the steam, and shut her eyes as they went by that door. Across the hall, up the stairs to the grandfather clock on the half-landing. Tock, tock, it went; time beating slowly on. Shadows darted towards her bedroom, running away from the flickering light of the candle.

The linoleum was cold under the soles of her feet as she undressed, fumbling with buttons and tapes. Her breath came and went in clouds in front of her face.

"We'll forget about washing just this once," Annie said,

pulling the nightdress over Sarah's head. Would that be black tomorrow? "Say your prayers now, and hop into bed."

Through her fingers Sarah watched Annie pick up the cast-off clothes, fold them neatly, lay them out on the chair for morning. Her mind was blank, unable to remember a single prayer. Besides, how did she know that God was listening?

What happened when you died? Where did you go? Think of Heaven – all those people since Adam and Eve. Heaven must be enormous to hold them all. How could you find anyone you knew?

"I thought you said them aloud," Annie said suspiciously.

Sarah remembered saying her prayers at Mother's knee; for a moment saw Mother's eyes smiling down at her. A lump came into her throat. She shook her head.

"You can have the lamp tonight as a special treat until you go to sleep," Annie said. "I'll leave the door open into my room. Then you've only got to call if you want me in the night."

The hand on Sarah's cheek was warm and rough. Mother's hands had been cool and smooth. Gentle, loving hands. Don't think of Mother. Don't think.

But how do you stop the thoughts that lurk at the back of your head like the wolves behind the wardrobe, waiting to jump –?

The lamp on the dressing-table was reflected in the mirror; two lamps instead of one. Annie had turned the wick down as far as it would go. The light was pale, hardly there at all, but enough to keep the wolves in their lair behind the wardrobe.

When Sarah stirred the bedroom was in darkness. Strange noises, scratchings at the window, sent her blundering into Annie's room, sobbing.

"Where's Mother? I'm frightened, I want Mother."

"Ah lovey, come to Annie. What is it then? Annie'll look after you, Annie's here."

"There's somebody outside . . ."

"It's only the wind in the trees, and the rain."

"... somebody tapping on the window."

"It'll be the creeper come away from the wall. We'll get Willis to see to it in the morning."

"I don't want to go back to bed."

Annie rocked her gently in her arms, stroking her hair. "Well ... perhaps, just this once ..."

Annie was in her nightgown, ready for bed, her hair twisted into one thick plait. The curtains behind her moved gently to and fro in the draught as the candlelight wavered over her face.

"You've been crying," Sarah accused.

"I've been thinking of old times," Annie said. Tears shone on her cheeks. "She was a good mistress, your mother. There's not many would have done what she did for my family. All that and never a fuss. Lord love us, what's to become of us all now?"

She sighed as she climbed into bed beside Sarah and blew out the candle. "Let's hope Miss Frances can tell us when she comes home."

But Frances said little when she returned with her sisters next day. She went grim-faced round the house, talked to Willis the gardener, spent time in the kitchen with Annie and the maids. She insisted on putting Sarah to bed herself, making Sarah wash more thoroughly than ever Annie did and clean her teeth twice because, Frances said, the first time wasn't thorough enough. When Sarah knelt to say her prayers, Frances said, "Come and sit on my lap. I want to talk to you."

Sarah sat, quiet, nervous, afraid to break the silence. Her other two sisters, Gwen and Julia, might be six and eight years older but they were children still, like her. Frances was not. Last summer Frances had put up her hair and let down her skirts. She was adult now and far removed from the world Sarah inhabited.

Frances sighed at last and smiled sadly at Sarah. "Poor infant. It's worse for you than for any of us, isn't it?"

"Annie turned my yellow dress black." Sarah's voice shook.

"You can't wear yellow when you're in mourning," Frances said, but she sounded matter-of-fact, not shocked like Annie.

"I didn't want to wear it, Frances. I thought I could look at it sometimes."

"I'll make you another yellow dress. You must be brave, Sarah; about everything, not only your dress. We are lucky, you know. We have each other. If we stick together and look after each other we'll be all right." She held Sarah close to her. "I'll do my best to take Mother's place, but you've got to be a big girl now, not a baby any longer."

"I'll try," Sarah said, and knew that she could not, as she had been about to do, ask to sleep in Annie's bed again; not tonight nor any other night.

Chapter Two

Outside the French windows, beyond the glass-roofed verandah, the garden lay sad and unloved in its winter clothing. The ragged grass of the lawn stretched away to beds of leafless rose bushes. Hummocks of dormant plants sheltered in the border under the sandstone wall. Beyond the shrubs and trees of the lower garden, under clouds heavy with rain, rolling farmland disappeared into the mist.

The greyness of the January afternoon had seeped into the dining-room where the Purcell sisters waited. White-faced and swollen-eyed, the three youngest stood in front of the fire, holding hands for comfort, while Frances gave them their instructions.

"You're not to open your mouths, do you understand? Not once. If you don't agree with what I say you can tell me later, when they've gone. If we stick together we'll be all right. I can manage Mr. Mackenzie; it's the lawyer I don't trust."

Her face was pale above the ugly mourning clothes, her eyes clear. If she had shed tears since coming home she had done so in the secrecy of her bedroom. She stood straight-backed, head erect, the thick chestnut hair glowing like a halo round her face. She looked like Boadicea going into battle, Sarah thought, and for the first time was partly comforted.

"They're coming up the hill now," Annie said, setting a lamp on the table. "There. That makes it look a bit more cheerful." She counted the chairs, checked the fire, tried to smile at the girls. "We'll know where we are in a little while."

The fire whimpered. Tongues of blue flame licked round the logs, turning yellow and orange as they reached up to the chimney. Outside the window water dripped steadily from the verandah roof.

Voices sounded in the hall. The children were quiet, listening. Annie's voice, greeting the visitors. That of the Rector, Mr. Mackenzie; of Mrs. Mackenzie, his wife. The thin, creaky voice that must belong to the lawyer. And finally an unknown, young man's voice.

Annie held open the door. "Mr. Mackenzie," she announced, as if the Rector were the only person she considered worthy of introduction. The dining-room was suddenly, bewilderingly, filled with large people and loud voices.

The lawyer, unused to children, greeted the Purcells awkwardly. Frances faced him as the Queen of the Iceni might have faced a Roman captain, Julia tried to smile, Gwen began to sob. Only the fire prevented Sarah from backing away altogether as the lawyer advanced towards her like a threatening black bird. His hand, when he shook hers, was limp and cold.

"You know my eldest son, of course," Mr. Mackenzie said. "Gabriel."

He was taller than his father, fair-haired like his mother, long-haired. Though Sarah had seen him singing in the church choir on his visits home from Cambridge, she had never done more than smile shyly at him after the service.

"By sight only," Gabriel said, holding out his hand to Frances. "Please forgive me. I know it's impertinent to come uninvited today but I thought I might be able to help. I thought I should know what's going on for my father's sake, but if you'd rather I didn't you have only to say and I'll leave at once."

Frances, standing stiff and unsmiling, relaxed visibly. "Of course I wouldn't ... it's kind of you. Won't you sit down?"

"I think Sarah should be taken out to Annie, Frances," Mr. Mackenzie said. "She's very young; there's no need for her to stay."

Sarah thought longingly of the kitchen, saw herself making toast with Annie in front of the range, remembered the soothing motion of Annie's rocking-chair, but knew what she had to say. "I must stay here," she said, trying to keep her voice steady. "Mustn't I, Frances?"

The lawyer bent down until Sarah could see herself reflected in his spectacles, her face pulled out of shape by the curved glass of the lenses.

"Come, come," he said, "this isn't an occasion for little girls. It has nothing to do with you."

Frances looked at him with scorn. "Indeed? Why not, may I ask? It's her future we're discussing as well as our own. Of course she'll stay."

He hesitated. "I suppose ... if she behaves."

"I'll see she's quiet," Gabriel said. "She can sit on my lap. Come on – Sarah, isn't it? Up you get."

She climbed up at once before he could change his mind. His knees were sharp and knobbly, quite unlike those of Annie or Mother, but he pulled her up over them until she was leaning against his chest.

"Is your name really Gabriel? Like the Angel?"

"Yes." He smiled across the table towards his mother. "Not that it's always very appropriate. Now hush. If we're not quiet we shall be asked to leave the room. We don't want that, do we?"

The lawyer sat at the head of the table. He cleared his throat, tugged at his cuffs, avoided people's eyes. He talked at length, about informal meetings, getting to know each other ... about probate and the time it took to deal with legal matters ... about executors, trustees, guardians ...

What did it mean? What did it matter, not having any relatives? They had each other.

"Of course if it is true that your mother ran away from home before her marriage and was disowned as a result, the lack of relatives is hardly surprising. Were you aware ...?"

Frances stared blankly at the lawyer. "Mother never talked of the past."

"Unfortunately your aunts — great-aunts, I should say — are quite unable to take on the care of four young girls. The elder Miss Ellison is bedridden and the younger, though anxious to be of help, is very old. As presumably you know."

"We've never met," Frances said. "They wrote to Mother every Christmas, but I didn't know we were related until Mr. Mackenzie told me so last night."

It was growing dark outside, the winter afternoon closing in. Soon Annie would come in to pull the curtains and shut out the winter night. Naval pensions, children's allowances, likely income — did Frances understand what he was talking about?

"... once the house has been sold and the proceeds invested we shall have a more accurate idea of your future income ..."

Julia's voice was high-pitched, loud. "Sell Hillcrest?"

"I don't foresee any difficulty. It's a pleasant enough house, so far as I can tell, and Somerset is a popular county. Professional men appear to favour the West Country when they retire from London. As do those from further afield," he added with a slight bow towards Mrs. Mackenzie.

Gwen burst into tears. "I don't want to leave my garden."

Frances leant forward. "Where do you expect us to live?"

The lawyer, taken aback by Gwen's tears, hesitated. "We assumed ... in Taunton. The trustees would rent a house and put a housekeeper in charge."

"Why?" Frances said.

He stared at her. "I don't understand?"

"Why move to Taunton? Why can't we stay here?"

"My dear Miss Purcell! Four girls, alone, in a house this size! It wouldn't do."

There were spots of colour on Frances's cheeks. Her hands gripped the table, skin stretched white over the knuckles. "But we don't want to live in Taunton. We want to stay here. Hillcrest is our home. It's the only home we've ever had — we always came back here for Father's leaves or when he went abroad. And when he died we came back for good ... You can't sell Hillcrest just like that. It's ours, all ours, everything in it. Everything. Every tree, every flower. Father planted the peach tree on his last leave; the maple to mark Sarah's birthday. You're just a lawyer," she cried passionately, "you don't understand things like that. You spend all your time with dry old bits of paper. What do you care about people or houses? Don't you see — it's all we've got left. We couldn't live in Taunton. It'd be like ... like being in a prison. You couldn't do it. You couldn't be so cruel."

The lawyer gave an embarrassed cough. "Come, come, Miss Purcell. Quite apart from the priority of the arrangement, there is the financial aspect to consider. It's a large house and the garden must be — what? Two, three acres? There's the running of it, the upkeep, the expense, Miss Purcell."

"There'd be expenses in Taunton," Frances said. She was calmer, quieter. "Rent. Food. A housekeeper, you said — her wages. If we stayed here we could keep Annie on. There'd be no rent. We grow most of the food we need. If the garden's too big we could let some of it grow wild, or have more orchard. Rent some out to Mr. Escott for his pigs. I don't see why size should be a problem. It doesn't cost any more. You don't know us. We aren't extravagant, we don't want expensive things. We want to stay here."

"It would need careful consideration. Mr. Mackenzie and I ... "

"Ah yes," Frances said. "Mr. Mackenzie. Of course, I know you're our guardian too, but Mr. Mackenzie's the one Mother wanted, isn't he, the one she chose. We should be here, near Mr. Mackenzie, not seven miles away in Taunton."

"There is something in what Frances says," Mr. Mackenzie

said. "I know that in our discussions it never occurred to either of us to keep the house on but perhaps ..."

"Please," Frances said to Mr. Mackenzie. "It does mean everything to us. We couldn't bear to leave Hillcrest, any of us."

"What about schooling?" the lawyer said. "I have already spoken to the headmistress of a suitable establishment in Taunton. She hopes to be able to fit you in at the beginning of next term."

"I'm not going back to school," Frances said. "Ever."

"There is no legal requirement for you to do so," the lawyer said, "nor Julia, who is – fifteen, is it? However, it would seem desirable for Gwendoline to attend for another year at least and of course the youngest child has her whole school life in front of her. What of *her*, Miss Purcell? Have you considered her?"

Frances drooped. "I'd forgotten about school."

"There is the village school," the lawyer suggested slowly.

Frances shook her head. "Not for Sarah. Mother always said she was clever. She's been reading for years for one thing. The village school wouldn't be good enough for her."

There was silence. In the grate a log slipped. The flames died down before flaring up again with a hiss. Gwen and Julia glared at Sarah as if it were her fault.

At last the Rector said, "I suppose I could ... I taught my own daughter. I still teach my youngest son at home. There is no reason why the girls should not come to me."

Sarah stared at the Rector with fear. It wasn't that she was frightened of Antony, the Mackenzies' youngest son, at least not exactly, but ... He was always trying to get her into trouble. In church he tried to make her laugh, turning round in the Rectory pew when Mrs. Mackenzie wasn't looking, and pulling faces at Sarah sitting behind. On one terrible occasion he had succeeded, making her giggle at the worst possible moment in the middle of the gospel reading. On another occasion, during a social meeting at the Manor, Sarah had innocently shaken his outstretched hand to discover, with a shriek of horror, that he had secreted a worm in his palm.

But if they were to stay at Hillcrest she would have to face Antony. Perhaps Mr. Mackenzie would help; he was always kind and sympathetic. He called regularly at Hillcrest in his role of Rector. His visits had frequently lasted longer than duty required for both he and Mrs. Purcell enjoyed a good discussion. "So refreshing to find a woman of independent thought," Sarah had once heard him tell her mother.

"You ought to talk to my son," he said when Mrs. Purcell talked of someone called William Morris whom she had known before her marriage. "You have much in common — he's something of a radical too." Mr. Mackenzie's voice warmed when he talked of his eldest son, his eyes lit up. "He's joined the Fabians, you know. I'm not sure that I approve, and his mother, of course ..."

"I can imagine," Mrs. Purcell had said and laughed. Mrs. Purcell had never cared for Mrs. Mackenzie. "Heaven preserve me from managing women," she would say, and disappear to the bottom of the garden if she thought Mrs. Mackenzie were likely to call.

Sarah thought Mrs. Mackenzie was wonderful; she could happily have sat for hours watching her, admiring her dresses, her hats, the way she sat, the way she moved.

Gabriel shifted in his chair, interrupting Sarah's thoughts. She sat up and rubbed her cheek where a button of his jacket had pressed into her flesh.

"All right?" he said in her ear. "I'm not used to little girls; you must tell me if you're uncomfortable."

She nodded, surprised to see Annie standing in the doorway. Annie avoided Sarah's eyes. Her fingers folded and unfolded the pleats of her apron as she glanced anxiously from Mr. Mackenzie to the lawyer and back again.

"She's very young," the lawyer said, as if Annie were deaf. "She can't be more than twenty-three or twenty-four. A housekeeper needs to be older, particularly in a situation like this."

"I shall be twenty-five next month, sir," Annie said quickly, "and I am used to responsibility."

Mrs. Mackenzie leant forward. "I know that Mrs. Purcell thought very highly of Annie's capabilities. And Frances is seventeen, old enough to run a home under supervision. I would consider it my duty ..."

The lawyer dismissed Annie with a brief nod and sat gazing into the distance. "We must do nothing in a hurry." His fingers tapped the table. "We shall have to consider. Now, if that is all ..."

"There is something else," Frances said. For the first time she sounded uncertain. "Would there be ... is there enough money for me to go to art school?"

There was a moment's silence before Mrs. Mackenzie said, "Art school? *Art school*?"

"It's hardly a question of money, Miss Purcell," the lawyer said, "but propriety. No respectably brought up young girl could attend – would want to attend – an art school."

Frances stared mutinously at him for a moment before looking at the faces round the table, one by one.

"What about you?" she said when her eyes reached Gabriel. "You haven't said a word so far this afternoon. What do you think about my going to art school?"

"I don't know. Art school – it's a matter of talent, isn't it? I don't know – are you any good?"

She tilted her chin in the air. "I'm not as good as I'd like to be at the moment but, yes, I have got talent."

He smiled. "All right. If you really want my opinion I think you should go – provided you can win a place at a decent one."

"Gabriel!" exclaimed Mrs. Mackenzie.

"Oh, come, Mother," Gabriel said. "We're not living in the dark ages now. There are some very gifted women about today. What about talent? Father believes in the parable, don't you, Father?"

"Yes, of course," Mr. Mackenzie said, but his voice lacked conviction. "Still – art school. Though I believe the one in Taunton is quite respectable."

"Taunton!" Frances said. "I wouldn't be seen dead in an art school in Taunton. I was thinking of London."

Gabriel's lips twitched. "I see. May I ask you how you intend to run a house this size and look after your sisters, while you are more than a hundred miles away in London?"

"It is impossible, Miss Purcell," the lawyer said. "Your mother may have bequeathed you her paints but I know that she would never have countenanced such an idea."

Frances appealed to Gabriel again. "I don't have to go at once. If I worked at home for a couple of years Julia would be seventeen. She could take charge during term-time, couldn't she?"

"That would be much more sensible," Gabriel said. "It would be a mistake to do anything in too much of a hurry. Give yourself – and everyone else – time to consider. I could see what I can find out about the various schools, if you like."

Outside the rain had set in for the evening, sounding a faint but continuous drum roll on the verandah roof. It was warm in the dining-room, Gabriel's chest comfortable to lean against ...

Sounds of departure woke her; the lawyer saying good-bye, doors opening, Annie offering coats.

"Very touching," Mrs. Mackenzie's voice said in the air above Sarah's head. "Let the poor child wake in her own time. You can follow us later."

The front door closed. Footsteps sounded muffled on the porch step. Gwen sniffed, about to cry again. Wooden rings rattled along the pole – Annie pulling the curtains across the windows.

Sarah kept her eyes shut, pretending to be asleep, hoping to stay on Gabriel's lap with Gabriel's arms round her for ever.

"I should have talked to you last night, Annie," Frances

said. "Perhaps you don't want to stay, now that my mother's no longer here?"

Annie's voice was surprised, hurt. "I wouldn't want to leave Miss Sarah, not at a time like this. I thought ... do you want me to go, Miss Frances?"

"No."

"I expect we'll manage all right," Annie said, but her voice was uncertain.

"We must manage," Frances said. "Make no mistake about that. If things don't work out here that man'll sell Hillcrest and settle the four of us in Taunton. Just like that. You know what that means, don't you, Annie? You'll be out of work. Without a reference from Mother too."

There was silence. At last Annie said, "I'll get the tea," and went out, shutting the door firmly behind her.

"That wasn't very kind," Julia protested. "Annie won't want to be out of work any more than we want to live in Taunton. Less — she's got that sick mother and her brothers and sisters to look after."

"Oh well. You know Annie and me ... at least she knows what's what now. I don't suppose they'll let us stay here, anyway."

"It sounded to me as if they intend to consider it," Gabriel said. "Tell me, how do I wake your sister?"

"Can't you let her sleep on for a bit?" Frances said. "She's been having bad nights, Annie says. I expect she's tired out. You'd better stay to tea."

The invitation was so grudging that Sarah would have added her own had she not remembered, just in time, that she was supposed to be asleep.

"Are you angry with me?" Gabriel said.

"Angry?" Frances said. "Why should I be angry?"

"I don't know. For saying what I did about London, perhaps."

"It was a bit tactless, wasn't it? You didn't have to say it. You could have kept quiet."

"So could you. What on earth made you bring up that art school idea? You were doing so well, you'd almost persuaded them to let you stay here and then you go and spoil everything. You're obviously not stupid; you must have realised the sort of reaction you would get from a fellow like that lawyer."

"It was important."

"Not that important, surely?"

"Yes, it was."

Sarah recognised the tone of voice. Obstinate; not to be argued with. It silenced Gabriel who waited several minutes before he said diffidently, "May I give you some advice?"

"If you want to. I don't promise to take it."

"Take things slowly. You all want to stay here apparently; I think you should concentrate on that. If you get them to agree and things work out satisfactorily, and they can see that they do, then start mentioning art school. Not ... how shall I put it? ... belligerently. Just mention it occasionally until the idea grows on them. Do try to look at it from my father's point of view, Frances. It's a frightening responsibility and quite unexpected."

"Oh, I know. I do know. I feel very badly about it. I don't know what Mother was thinking about asking him to take us on. Why didn't she consider it before? Why didn't she make a will when Father died? She was such a practical person usually. To leave it until she was dying herself ... I suppose she thought that there'd always be time. And she was quite all right when we went back to school last week, you know, apart from getting caught in the rain the day before. I still can't believe it."

"I'm sorry," Gabriel said. "I didn't mean to upset you."

Frances blew her nose. "It's all right. You will stay to tea, won't you?"

"Well ... if my parents are discussing matters with that man, I think I should be present to put forward your views. Perhaps you could invite me another day?"

It was time to wake. Sarah stirred, opened her eyes, gave what she hoped was a convincing yawn.

"She's waking up at last," Gabriel said, and smiled down at her.

She tried to sit up but was caught by her hair which had twisted itself round the buttons of his jacket. She let him unwind the strands and set her down on the floor but when he tried to straighten her pinafore she was overcome by shyness and scuttled across the room to grab Frances's hand.

"You look like a little harvest mouse hiding behind your sister's skirts like that," Gabriel teased.

"You will try and persuade your parents to let us stay here, won't you?" Frances said, still holding Sarah's hand as she went with him into the hall.

"Yes, of course. I suspect that it'll come down to finance in the end: whether you can afford to live here or not. It's a nice house, isn't it? I can understand why you want to stay."

Frances watched him put on his coat. She said hesitantly, holding open the front door, "Did you mean what you said about finding out about art schools? I do need to know. I was thinking of the Slade – it sounds the best place for me – but I'm not sure how you get in. I wrote and asked but we were only allowed to get letters from relatives at school and I think the headmistress must have confiscated their reply. I never got one, anyway."

"I'll see what I can do. There shouldn't be any difficulty once I get back to Cambridge. I often go down to London. I'll write and let you know. Think over what I said though. I wouldn't talk about it too much if I were you. There's plenty of time." He added impulsively, "I had no idea you were such a determined person. I'm sure you'll go far."

"Of course I shall," Frances said. She sounded surprised, as though it had never occurred to her that she might do otherwise.

Gabriel laughed as he set off down the hill. His words floated back through the rain to the two girls standing in the lighted doorway.

"Of course you will. But please don't forget to invite me to tea before you leave."

1913

Chapter Three

Today was Friday, Sarah's favourite day, waited and longed for throughout the week. It was very far from Annie's favourite day, however, although it was her day off, for on Fridays Annie trudged the six miles to her old home under the Blackdown Hills where she cleaned the cottage, cooked, washed and ironed for her bedridden mother and the brothers and sisters who still lived there, before wearily trudging back in the evening to Hillcrest.

Mrs. Mackenzie had been shocked to discover how much free time Annie had been allowed by Mrs. Purcell. She had said so – at length, for Mrs. Mackenzie was not one to mince her words. She talked about the undesirability of giving one's servants too easy a time, the dangers of setting unfortunate precedents, the evil of providing bad examples to others in the village; and when the Purcells stared at her, she told them that with her experience in domestic affairs they would be foolish indeed to ignore her advice.

She could have said nothing more unfortunate as far as Frances was concerned. Frances tilted her chin in the air and told Mrs. Mackenzie that she intended carrying on as her mother had done. Her mother had allowed Annie to return home one day a week and so would she. Mrs. Mackenzie should know that Frances didn't give a fig about precedents, examples or what other people thought. There were many things Frances didn't give a fig about: regrettably Mrs. Mackenzie's opinion was one of them.

Mrs. Mackenzie had come to accept Annie's time off in the end – she knew Annie's worth, after all – although the subject was still brought up from time to time, particularly when Mrs. Mackenzie was more than usually irritated by Frances.

"She's been going on about Fridays again, Annie," Frances said one day when Sarah was in the kitchen helping Annie break blocks of salt into powder. "You know I don't mind, but you're only their half-sister and I can't help thinking that they take advantage of you. They'd have to manage if you were the other side of the country. Isn't it about time the boys were thinking of marriage?"

"It's not that easy, Miss Frances. You've met my mother – not that I blame her, mind, there's no knowing what I'd be like if I was in pain all the time – but a girl'd have to be pretty desperate to take her on as well as a husband, wouldn't she? I'm sorry, really I am. I don't want to make trouble between you and Mrs. Mackenzie."

"I shouldn't let that worry you, Annie. We're like oil and water, Mrs. Mackenzie and I. We'll always fight."

"That's no way to talk, Miss Frances," Annie said severely. "A pretty pickle we'd have been in if Mrs. Mackenzie hadn't been around when your mother died."

"You know we were a godsend to her – another household to set about. Well, she can say what she likes to me but I'm not having her organising you. You have a much harder time than Bertha or Hilda – or any of the Rectory servants, come to that. I sometimes think taking girls straight from school is worse than having no servants at all. And you know that hatchet-faced housekeeper at the Manor always says how well you've trained the girls who go on from here."

It was Annie's absence that made Fridays so special. To make things easier at Hillcrest while Annie was away Sarah stayed on at the Rectory after her lessons had ended and lunched with the Mackenzies. The custom continued long after the need for it had gone. Sarah hoped that it always would, for lunch at the Rectory without her three sisters was her idea of heaven. On her own she could pretend to be one of the Mackenzies, part of a proper family at last, with her own mother and father. Every child she knew possessed at least one parent, usually both, and aunts, uncles and often grand-parents too. There was no-one else in the village whom the Shattock boy could shout 'Orphan' at in such horrid, jeering tones. Sarah hated being different and the Purcells were different in so many ways. That was why she wanted to go away to school. At boarding-school her lack of parents, the peculiarities and unconventional behaviour of her sisters would be unknown. At boarding-school she would be like everyone else.

She had experienced in the past the disappointment of unanswered prayer, and although she knew that fulfilment of this particular desire was improbable she prayed nightly nevertheless. Perhaps a flash of light would make Frances, like Saul on the road to Damascus, understand how necessary it was ... In the meantime Sarah, who spent the greater part of her life living in dreams of one kind or another, created her own fantasy world of school in exercise books that she hid in the back of her stocking drawer, away from amused or prying eyes.

But today was Friday. Sitting at the Rectory table in the familiar Rectory dining-room, Sarah was pretending to be a true daughter of the house, telling herself that Antony, sitting opposite, fair hair flopping over his eyes in imitation of his eldest brother, really was the brother he so often seemed to be, and Lucy, her elder sister; that Mr. and Mrs. Mackenzie had become her father and mother.

It was easy for the Rector to take the place of Commander Purcell, by now only a shadowy figure in Sarah's mind; more difficult for Mrs. Mackenzie to replace the still vivid memory of Mrs. Purcell. Mrs. Mackenzie was regal rather than motherly and it was admiration rather than love that Sarah felt for her. Long ago Antony had dubbed his mother the White Queen. Mrs. Mackenzie's deportment, her beauty and elegance, the slight air of remoteness, together with the way in which she ruled Rectory and village, made the name not inappropriate. Mr. Mackenzie might be responsible for the spiritual life of Huish Priory but it was Mrs. Mackenzie who saw to the rest, running village and villagers as calmly and efficiently as she did the Rectory and her family.

It was the clockwork precision of life in the Rectory that delighted Sarah and never ceased to amaze her; the way in which everything happened as and when decreed. Lunch was served at a quarter past one and at a quarter past one every day, not a moment before, not a moment after, there it was on the table. At Hillcrest, meal-times depended on how the painting was going, or the jam-making, on whether the sun was shining or not, on a hundred and one little things that would never have been allowed to disturb the smooth running of Rectory life.

There was a bitter-sweetness about lunch today. It was the last Friday of term, the last time Sarah could expect to be here without her sisters for several weeks, and just because she wanted the meal to last for ever it was over in no time at all, with Mr. and Mrs. Mackenzie retiring as they did every Friday, she to rest on her bed, he to read the *Church Times* in his study.

Antony tipped back his chair and yawned. "Ah, well. The holidays are here again. Geoffrey this afternoon, Gabriel next week – the Mackenzie family reunited once more. Back to the mad social whirl. How's your tennis coming along, Sarah?"

"It isn't. I haven't anyone to practise with. They're all too busy."

"That doesn't matter. There are masses of walls at Hillcrest you can use."

"They've got things growing over them."

"Try bashing the ball between branches. You couldn't have better practice than that."

"You'll get her into trouble," Lucy said. "Imagine Frances's fury if she found the fruit flattened against the wall."

"Frances wouldn't mind," Antony said. "Knowing Frances she'd go into ecstasies over the colours and paint portraits of squishy pears for days on end. Wouldn't she, Sarah?"

"I expect so." She added wistfully, "You did say you'd teach me tennis this year, Antony."

"So I did. Never mind, it's nice for us to have a ballboy. I'll teach you next year. Really I will."

"I'll spend a few minutes with you, Sarah," Lucy said as she went out of the room. "Later, when I've finished my practising."

In the sunlight outside the dining-room the tennis net sagged across the grass court. Antony made no move nor any suggestion for occupying the afternoon. He was waiting for his brother Geoffrey. Now that the holidays were here he would desert her and join the others. Sarah could not blame him. He was five years older than she and although in term time he was still glad of her company when the holidays came he naturally preferred that of the adults. And her village friend, Jess Hancock, released from school, would be expected to look after her younger brother and sister. The holidays stretched out into the future, empty and lonely.

She stood disconsolately on the Rectory doorstep. The notes of the piano, played with the soft pedal down, floated out from the drawing-room. "Loving Shepherd of Thy Sheep." That was what they would be singing in Sunday School this week. She fingered the halfpenny in the pocket of her smock. Perhaps one of Mr. Cross's treacle toffees would cheer her up.

Mr. Cross's bad tooth winked blackly at her. "And what's Master Antony sent you to buy today? Another tin of elbow grease? Pigeon's milk? Oh, he's a rare one, is Master Antony."

Sarah reddened. "I'm not going to shop for him any more."

"So your sister's home again," Mr. Cross said, weighing out the treacle toffees. "For good this time, is she? Funny do, when you think about it. How many pictures has she sold? Eh?"

Sarah avoided his eyes. "Lots."

It was hot outside the shop, too hot for tennis. The air smelt of heat and dust. Nothing moved in the village street. Even the schoolhouse was without life. Its windows, set high above the ground, allowed light but no distraction. Inside, Jess, far in advance of her contemporaries in the classroom, would be helping Miss Ross teach the little ones. Sarah had once suggested that Jess should join her for lessons at the Rectory, and had been surprised by everyone's astonishment.

She walked slowly past the wrought iron gates of the Manor. Drawn blinds made the house look uninhabited and unloved. A peacock dragged its tail wearily across the grass towards the peahens dozing under the trees.

The road became steeper as it passed the entrance to the Rectory. Sarah glimpsed the Rectory itself through the trees and hesitated for a moment, wondering whether Lucy would have finished her practising, before walking on in the shade of the chestnut trees that separated the churchyard from the road.

She paused as she always did when she reached the brow of the hill and Hillcrest. Hillcrest gave her a warm, comfortable feeling. It was solid, permanent. It was home. Yet Hillcrest was like the Purcells themselves. It stood slightly apart, did not belong. It had turned its back on the village and looked away instead, over rolling farmland to the grey outline of distant hills.

She smelt the jam as soon as she opened the front door, sticky, sweet and hot. The kitchen window, shut against wasps, was wet with condensation. Gwen's flushed face glistened as she bent over the jampot labels.

"You *have* made a mess," Sarah said. "Why've you spilt it over the table? It's on the floor too. Annie'll be furious."

It didn't matter what you said to Gwen. You had to be careful with Frances but Gwen was never cross. She said now, "The wretched stuff boiled over when I wasn't looking. Get a cloth, there's a good girl."

"Why doesn't Frances want to sell her paintings?" Sarah said as she went over to the sink. That was something else about Gwen: she would always listen and try to answer your questions.

"She doesn't think they're good enough yet. She has very high standards, you know. I don't think she likes the idea of her early, not-so-good work going round later on." She looked up from the labels. "And talking of work, Sarah — Lucy had a word with me the other day. She says you haven't been learning the collect properly for Sunday School."

Sarah carried the hot jars from the table to the dresser with great care. Should she arrange them in threes or fours? Or fives, making them easier to count? "I think it's mean," she said at last. "Her telling you things like that, just because you're a friend. She doesn't tell Mrs. Hancock when Jess forgets."

"She knows it wouldn't do any good, I expect. She said you read it off from the book in your lap last Sunday."

"I didn't think she'd noticed," Sarah said, pink-cheeked, and added quickly, "It wasn't my fault. I didn't have time. I had to carry Antony's things when he went hunting butterflies."

Gwen raised her eyebrows. "I can imagine what Lucy had to say if that was the excuse you gave. I don't know why you make such a fuss. You learn great chunks of Latin and Greek without any trouble. Go and get the prayer book. You can start on this week's now."

"Don't you want me to help clear up?"

It was unkind to take advantage of anyone who could so easily be distracted. The mess, apart from the treacly stuff on the range which poor Elsie would have to clean up in the morning, was not as bad as it looked. In no time at all she and Gwen were sitting at the kitchen table . . .

It might be wonderful to live at the Rectory as a member of a proper family, Sarah thought, but there would be disadvantages. Cook ruled the kitchen. Mrs. Mackenzie excepted, no-one from upstairs was allowed over the threshhold. Cook would never let you scrape jam off the table and lick the knife clean afterwards, whereas Gwen, all thought of Sunday School and the hated collect forgotten, merely giggled and said, "Oh Sarah, honestly! What would the others say?"

Chapter Four

It was Geoffrey who came to Sarah's rescue. She was so relieved to see him, standing by the gate into the vegetable garden blinking in the sunlight in the awkward manner that was somehow characteristic of him, that she nearly burst into tears.

He waved in reply to her cries and came over to the fruitcage. "Gwen said you were in the garden. Are you pinching the raspberries by any chance? I say, you've got a bird in there."

"I know. It's all my fault. They'll be furious. I tried getting it out but it won't go and I've gone and made a hole in my stockings and they were new on today and Annie'll be cross too and she'll . . ."

"I say, old thing, calm down. There's no need to cry. We'll get him out in no time at all. Look, you go up that row and I'll go up this. We'll trap him in the corner."

The long arms of raspberry canes snatched at her smock as she advanced with thumping heart between the rows, driving the bird – a young thrush, quite as frightened as she was – in front of her. It squawked and flapped until, trapped at last, it plunged into the netting and tried to strangle itself with the oiled black rope. Geoffrey bent down and placed his hands gently round the throbbing body.

"Poor little thing. He was frightened, that was all. Come and look. He's beautiful, isn't he? Go on, touch him. He won't hurt you."

She stroked the bird nervously, standing as far away as she could. The feathers were soft and warm under her fingers. Its head peered out above Geoffrey's hands, the beady eyes watching her.

"We'd better set him free, I suppose," Geoffrey said, and going outside the fruitcage, he launched the thrush into the air. It swooped joyfully up towards the sun. "You must have a pretty big hole to let a bird that size get through," he said.

"It was the gate," Sarah said guiltily. "I must have left it open this morning. You won't tell them, will you?"

He grinned at her. "What's it worth?"

She considered. "There are some strawberries left if you look carefully."

"I was only joking," he said. "I wouldn't say no to strawberries, all the same."

There was a surprising number of berries hiding under the dense round leaves. Sarah and Geoffrey gathered them into a pile and sat sucking the fruit from the stalks in a manner which would have been frowned upon by Annie as well as Mrs. Mackenzie.

It was very warm. Nothing moved. Willis had gone home long ago. Sarah looked anxiously at Geoffrey, afraid that he might be bored by her company. Perhaps he wanted to go home too but was too polite to say so. He seemed quite happy however, content to sit in companionable silence in the sun. His hands were too big for the rest of his body; they hung over his bent knees as if he were not sure where to put them. Unlike the rest of the family, who had inherited Mrs. Mackenzie's fair good looks, he was as dark as his father had once been, with soft brown eyes and disorganised eyebrows that overlapped above his nose and gave him an undeserved sulky appearance.

"Gwen said the others have gone over to Dunkery."

"They went to see Jane Lynch. She wanted Frances's advice about her watercolours."

"She's laying herself open a bit, isn't she? Her painting's worse than Lucy's." He sounded amused as he lay back on the straw, hands under his head. "Oh, it is nice to be home. You don't know what a relief it is to have wiped the dust of that place off my feet."

"I can't think why you hated school so much." It seemed very unfair to her that he could not appreciate something she so desperately wanted.

"You would if you'd been there. Mind you, Cambridge will be as bad. Think of carrying on with that awful Latin and Greek for three more years. And what good is it all in the end? Actually, it isn't just Cambridge." He sat up. "It's Gabriel, too. Oh Sarah, it's so awful being the middle one, you can't imagine. Sometimes I wish I'd never been born."

"But . . ."

"I'm so ordinary, you see, and there's Gabriel ahead of me who – well, you know what Gabriel's like. And now Antony's coming on and he's so clever he jabbered Virgil in his cradle . . ."

She was awe-struck. "Did he really?"

Geoffrey gave her an unexpected, affectionate smile. "You are funny. I was only joking. Not that I'd have been surprised if he had. And it's not a joke either. I'm not stupid, you know, Sarah. I did quite well at school but they all expected me to be as good as Gabriel and I'm not. He was so good at everything, not just lessons – rugger, running, and marvellous at cricket. He managed to hit the school greenhouse with a cricket ball, you know. No-one else has ever done that, though lots have tried. It'll be the same at Cambridge. Gabriel got a first and now he's a Fellow. How can I follow that?"

"Oh, Geoffrey . . ."

"Mother thinks he's wonderful. Father says Antony is. No-one thinks anything of me." He frowned down at his feet. "Everyone goes on about Gabriel so. He's not all that marvellous. He can be awfully bad-tempered sometimes. People don't see that, though. They think he's clever and – would you say he was good-looking, Sarah?"

When she had been younger and still reading fairy tales she had always seen Gabriel in place of the prince. "Yes, I would."

"Would you say that I was?"

She looked at his face, kind, comfortable, unhappy. "Well . . ."

He hunched his shoulders. "You don't have to say. I know. I don't suppose any girl will ever look at me. They all adore Gabriel."

Why should she hate him saying that? "How do you know?"

"He's always getting letters from girls. You should see Mother eyeing the writing on the envelopes. As a matter of fact, I don't think he cares. I think – you won't tell anyone will you?"

"Is it a secret?"

"Yes."

She loved secrets but rarely had any. "I won't tell."

"I think he's interested in Frances."

She stared. "Frances? Frances? What makes you say that?"

"When you see them together . . . and it's very odd the way he spends so much time at home nowadays. He never used to. He was always off on Fabian summer schools or campaigning against the Poor Law or doing things like that."

"He's away now."

"He'll be back on Tuesday. Any sensible person would stay in Germany for a couple of months once they got there, but Gabriel comes back after a couple of weeks. I'm sure it's because of her."

"Do you mean he's going to . . . well, is he going to marry her?"

"Mother keeps saying he should settle down. He's twenty-four, you know. She says he doesn't meet the right sort of girl in Cambridge. They're all wild or intellectual, according to her." He said thoughtfully, "Though I don't suppose she'd consider Frances the right sort of girl either."

Sarah flushed. "Why not?"

He said awkwardly, "I didn't mean ... oh, you know Mother and Frances. They do rub each other up the wrong way. And Frances is ... different. Most girls don't paint like she does. Seriously, I mean. She always says what she thinks, too. I'm quite frightened of her sometimes. I do like her, of course," he said hastily, "but you never quite know what she's going to say or do next."

Sarah stared at the green mound of strawberry plants stretching away to the edge of the cage. She didn't know what to say. "I wish you hadn't told me," she said at last.

"I'm sorry." He wiped his hands down the side of his trousers. "Forget it then. It's probably not true, anyway." He unfolded himself and stood up, catching his hair in the rope of the overhead netting. "Tell Julia I'm home," he said when he had freed himself. He gave her an awkward flap of one hand as he departed, leaving her alone in the fruitcage.

Marriage.

Her first thought, heart pumping with sudden panic, had been: what will happen to us? How will we manage without Frances?

For the first year after Mrs. Purcell's death there had been a real fear that they might be made to live in Taunton. It was never talked about in front of Sarah but she was aware of it nevertheless, like a heavy black cloud pressing down on them all. Sometimes, when she had been naughty or if she woke suddenly in the middle of the night, it recurred: panic at the idea of her secure world suddenly made insecure, of banishment from the Garden of Eden. Such fear had long ago faded but it came back now, as sharp, as frightening as it had ever been.

If Frances married and went away would we be allowed to stay?

She told herself not to be silly. Frances had spent every term for the past three years in London. The rest of the Purcells had managed very well without her. Julia didn't mind running the household, Gwen actually liked it. And when she was at home Frances spent all her time up in the stables, painting.

Perhaps it would be all right. Talking of marriage didn't mean ... Bill Roberts and Jess's sister, Mary, had been walking out for two years; they didn't intend to get married until next summer. That made a three-year engagement. Sarah cheered up. Anything could happen in three years. In three years she herself would be thirteen, really quite old and responsible.

All the same, she thought, hugging her knees, I wouldn't want Frances to belong to anyone else. Frances belonged to the Purcells, not to Gabriel Mackenzie.

Perhaps Geoffrey had imagined it. Perhaps he'd said it for something to say. Antony often made outrageous statements simply because he enjoyed the reaction such statements provoked. But Geoffrey wasn't like that. Geoffrey never tried to draw attention to himself. Partly because of his clumsiness which often irritated his mother almost beyond endurance, Geoffrey always tried to stay unnoticed in the background.

The afternoon's conversation had surprised Sarah. Of course she understood how diffficult it must be coming after someone like Gabriel – hadn't she the same problem, three times over? – but to dislike being the middle child ...

It was worse, much worse, being the youngest: never big enough, or old enough, or strong enough, always left out, left behind or forgotten, never taken seriously. And however much she longed to be grown-up she knew it would always be the same. For the rest of her life her sisters would be out there in the distance, six years, eight years, ten years further on.

It wasn't only that they thought they knew best, ordered her about, told her what to do, what to wear. There were other problems, of expectation, of comparison. Fairy Godmothers grow tired by the time the fourth child comes along, their gifts diminish. Beauty, talent – Sarah could see the decline.

First at the Purcell christenings had come Frances – not beautiful, Frances herself said, but people turned to look nevertheless and kept looking, however provoking she was being, admiring the glossy, chestnut hair, the laughing eyes, the impudent smile. And Frances was clever – a genius Gabriel had once said, and been reprimanded by his mother for saying so – putting herself through art school with prizes and scholarships.

People looked at Julia, too, though not in the same way. Julia was less striking, her painting less brilliant than that of Frances. Without Frances ahead of her, people would have admired and applauded. With Frances as an example they said, "Well, yes, of course – but what would you expect? It's in the family."

The Fairy Godmother was beginning to tire by the time the third christening came round: Gwen was nothing outstanding to look at (though better than me, Sarah thought) nor as original a painter. Her work was quiet, self-contained, like Gwen herself: intricate studies in pen and ink, delicate water-colours of flowers and plants.

Sarah's arrival was too much for poor Fairy Godmother, who had forgotten about beauty altogether. My eyes are too big, Sarah would think, studying her face in the mirror, and I've got freckles – freckles! – on my nose and it's all very well Annie saying it's my fault if I go out in the sun without a hat but I can't help forgetting sometimes. And my hair is just plain mouse which is why Gabriel calls me his mouse and it gets dreadfully tangled and I don't care if Annie says a hundred strokes night and morning, she doesn't know how long it takes or care how much it makes my arm ache.

But worst of all the Fairy Godmother's omissions was talent. Sarah was no artist. In a household where painting and drawing were as much a part of life as breathing, nothing could be worse.

"I can't understand it," Frances had once said. "Mother was so gifted, and the rest of us – you'd think there'd be a spark in you somewhere."

That was when Sarah decided never to draw or paint again. However much she might long for Frances's approval she knew that she would never be able to gain it in that direction.

Not until Geoffrey had returned to the Rectory did it occur to Sarah that Lucy was a middle child, too. Geoffrey must have forgotten Lucy. People did forget her, or at least not notice her, despite her work in the Sunday School and the Sewing Circle and the way she ran the Girls' Friendly Society. Lucy was part of the Rectory, taken for granted. Yet Lucy herself never took things for granted. It was Lucy who tried to see that Sarah was not always left out, who occasionally banged a tennis ball at her, who, when Sarah was ill, read to her for long hours on end. Did Lucy object to being in the middle, too, Sarah wondered, or did she accept it like everything else?

One of these days I'll tell Geoffrey what it's like being the youngest and then he'll be glad he's in the middle, Sarah promised herself as she sat looking at the familiar vegetable garden through the unfamiliar black diamonds of the netting; until the shadows of the elms fell onto the fruitcage and Annie, back from the Blackdown Hills, called her in to bed.

Chapter Five

Sarah's nose was in a book again. Lying on the verandah in the shade of the wisteria, the flagstones cold and hard under her stomach, she was reading *Micah Clarke*. Like most of the books she read it had once belonged to Gabriel, for ever since he had found her one afternoon sobbing in the shrubbery over *The Story of a Short Life*, Gabriel had kept her supplied with the stories he had liked as a boy. Henty, Harrison Ainsworth, Conan Doyle – there were books enough in the Rectory to entertain her, and Jess, too, for years to come.

"They're better for the child than Victorian tragedy," Sarah overheard Gabriel say to Frances. "You ought to keep an eye on her reading, you know. She'll be getting morbid if you're not careful."

Sarah felt warm and comfortable, as if she belonged, holding a book in her hands that Gabriel had once enjoyed, reading the name on the flyleaf written in strange, round letters so unlike his present writing. She would murmur the words over to herself, "Gabriel Edward Mackenzie".

Today the conversation around her, as her sisters and the Mackenzies rested after a game of croquet, was of more interest to Sarah than the story of Micah Clarke. Annie disapproved of eavesdroppers but Sarah told herself that eavesdropping didn't count as eavesdropping if people knew you were there. After all, no-one ever told her anything. Listening when others thought she was reading was often the only means of discovering what was happening, or about to happen.

"I do wish you wouldn't take your shoes and stockings off when you play, Frances," Gabriel was saying. "It unnerves me. It's the only reason I lose. I'm scared of hitting your toes."

"Excuses, excuses," Frances said, wriggling her toes and laughing. "Why don't you take your shoes off, too? It's much more comfortable. Go on, try it."

"You'd still win. You wouldn't worry about my poor feet – I wouldn't put it past you to try and cripple me on purpose."

"You're savages, the lot of you," Geoffrey said. "Still, I'd rather be with you than with the civilised mob at the Rectory. Mother's having another of her tennis parties this afternoon."

"I can't bear it," Gabriel said. "Where does she find the girls? None of them can hold an intelligent conversation. I've a good mind not to go home. I'll escape up into the Quantocks and not go back until dark."

"I must be there to see Mother's face," Antony said.

"We always said that one day we'd get up really early and see how much of the Quantocks we could explore."

"Why don't we go tomorrow?" Geoffrey said. "This afternoon won't seem so bad if we have that to look forward to."

Sarah sat up. "Can I come too? Please take me with you."

They turned and stared.

"Don't be silly. You couldn't keep up."

"I could. I'd run."

"You're too little."

"You're always promising to take me up to the Quantocks and you never do."

"Poor brat," Gabriel said. "She always gets left behind. Couldn't she come for once?"

"We certainly shan't get far if we have her with us," Frances said. "You'd be the first to complain, Gabriel, you know you would."

"We could carry her. Geoffrey and I could take it in turns. You wouldn't mind, would you, Geoffrey?"

Frances shrugged. "It's up to you. I think you're mad. And don't blame me if we get no further than Bagborough."

It was not only Frances who was against Sarah's joining the expedition. Julia and Gwen had their doubts too, and as for Annie . . .

"Mr Gabriel's taken leave of his senses," she said. "Ridiculous, that's what it is. How does he think you can keep up, I'd like to know."

"But you'll have such a peaceful day without me, Annie," Sarah said, trying to win her onto her side, terrified that the weight of opinion against Gabriel might persuade him to change his mind. "You'll be able to finish *The Sailor's Love-Knot* without one interruption."

"H'm. That's as may be. Mind you tell them when you're tired, that's all I can say. And remember you've got to get home again. No good saying you can't go another step when you're halfway along the ridge."

The world was still asleep when they set off next morning, the lane quiet and shadowy, protected from the light of the early morning sun by the height of its hedges. Leaves and grass were weighed down by dew. Beyond the closed iron gates of the Manor the peacocks remained silent and unmoving. Sarah walked softly down the hill behind the others, anxious not to attract attention, fearing that she might yet be sent home.

Past shuttered front parlours they went, past the forge where the fire was already throwing leaping shadows into the corners and Bill Roberts, stripped to the waist, waved a hand in greeting; past inn and orchards, and the cottage behind whose windows Nanny Mackenzie kept an eye on village comings and goings, until they were in true country at last.

"Lucy nearly wasn't allowed to come," Antony told Sarah,

stopping on the hump-backed bridge to throw a stick into the stream. "It's the Mothers' and Babies' Circle today. Or the sewing group. Something, anyway. Mother expected her to help."

"She is here though," Sarah said.

"Yes, well." He moved over to the other side of the bridge and watched the stick come out into the sunlight and float away towards the spinney. "Mother and Gabriel had a row about it at dinner last night. Gabriel won. He usually does if he's determined enough. He wears Mother down."

Lucy and Frances talked earnestly together as they walked in front. Perhaps Lucy was telling Frances about the row. Sarah watched and wondered. If what Geoffrey had said were true, wouldn't it be Gabriel rather than Lucy walking with Frances? Ever since that afternoon in the fruitcage Sarah had watched her sister and Gabriel, but was no wiser. What did people do if they were 'interested' in each other; if they were thinking of marriage? The stories she read did not tell her. She learnt nothing from seeing Mr. and Mrs. Mackenzie together. There was no-one she could ask without betraying her fears.

Antony moved away to join Gwen. Geoffrey and Julia were laughing together. Sarah stared at them. Fingers knotted in her stomach. Frances and Gabriel, Julia and Geoffrey, Gwen and Antony. She would be left at Hillcrest with Annie. Two of them, in Hillcrest. No, that would not be possible. They would have to move to Taunton. For a moment panic seized her. Perhaps she would be able to move into the Rectory with the Mackenzies. But Annie would leave rather than live at the Rectory and Sarah knew that she could never part from Annie.

"You're really grown-up now, aren't you?" Gabriel said, smiling down at her. "You must tell me if we're going too fast. We don't want to tire you out so early in the day. Frances tells me I always set a punishing pace."

His hand in hers banished her fears, for a little while at least. Gabriel would look after her. And it would be a pity to spoil the day by worrying over something that might never happen.

She was on familiar ground as far as the common. There the road divided into two. One road led to Dunkery St. Michael, a road Sarah knew well, and from there went on to Taunton. It was the other road they took today, heading north across the common. A gaggle of geese marched in formation towards them, stiff-necked, hissing.

"They think they're in Rome," Antony said and gave a nervous laugh. Sarah noticed him drawing closer to Gabriel.

"I do wish the Quantocks weren't so far," Gwen said. "It takes such an age to get there."

"What rubbish," Gabriel said. "It doesn't take long. You can't bear to leave the garden, that's your trouble."

It seemed very far to Sarah. She thought that they must have been walking for hours and hours. The road twisted round corners, climbed up, came down, went on for ever. There was no sign of the Quantocks. Perhaps they had been a mirage. Perhaps they had never existed.

They paused in gateways to admire the view and allow Sarah to rest. The countryside spread out to the horizon, fields of stooked corn, fields of stubble splashed with the green of clover, ploughed fields of wine-red earth.

"When I'm rich," Frances said, leaning on the top bar of the gate, "I shall build a hut at the bottom of the Quantocks to keep my paints in. Then I shan't have to carry them all this way."

"You haven't brought your paints today, have you?"

She shook her head. "Not paints. I always carry pencils and a notebook though."

They came to a village of thatched cottages, red sandstone walls, dark green rhododendrons. Crossing the railway reminded Gabriel of London. He told them about living near railway tracks, of conditions in large cities and families, respectable, working families, trying to live on less than a pound a week. Friends of his, he said, fellow Fabians, had been making a survey in Lambeth to see how such families existed, how the government might help. "They're women doing it, you know, some of the Women's Group. It's easier for them to talk to the wives."

"I believe you're trying to recruit Somerset women to the Fabians," Julia teased.

"You may laugh, but I'm serious. Imagine trying to bring up a family with two or three children on twenty shillings a week."

"They shouldn't have children."

"Frances! How can you say such a thing? Sometimes you shock me, really you do."

"Why? I shan't have any children. I couldn't, and paint as well. I'm going to paint. You have to choose."

"You can say that now." Gabriel sounded impatient. "You'll feel differently when the time comes."

"It would solve all your problems, though," Antony said. "No children – you'd get rid of the poor in one generation. Had you thought of that?"

"One of these days I shall tell your mother that you're a socialist," Frances said, her eyes mocking.

Gabriel was always talking about the Fabians, about poverty and the Poor Law, the need for changes in society, things that Sarah tried hard to understand for his sake but could not. Gabriel's hobbyhorse, Mrs. Mackenzie called it indulgently. She was not always indulgent. She and Mr. Mackenzie disliked Gabriel's talk of liberal reforms, considered that new legislation would sap people's initiative, that private charity was enough. Discussion at the Rectory dinner-table was sometimes far from restrained.

When people talked of poverty Sarah remembered *The Water Babies* and the little of Dickens that she had read. Poverty was in the past; children no longer went down mines or small boys up chimneys. Yet she knew that in winter Jess often had nothing more than potatoes to eat and that Annie, giving her extra food in the kitchen, would say that it was all very well seeing to the men in the family but she was sure the Hancock children were half-starved.

Hills took away the breath for arguments. They were climbing steeply now, trees and fields dropping away. The ridge of the Quantocks rose up in front of them, smooth and rounded against the sky. Sarah's breaths came in gasps and hurt against her side. The path twisted between bracken and gorse and came out at last onto grass, wiry tussocks of it, the soil black and peaty beneath. Sarah thought of rabbit holes. Like Micah's horse it would be easy to stumble.

Gabriel dropped back. "All right, mouse?"

She nodded. "*Micah Clarke*'s about the Quantocks," she said. "Well, Monmouth and the battle of Sedgemoor really, but the Quantocks come into it."

"We should have gone up the other way and shown you Cothelstone. Judge Jeffries had two of the rebels hanged on the gateposts there. The lord of the manor tried to persuade him to show mercy and that was Jeffries' way of teaching him a lesson."

She stared at him, wide-eyed. "Are they still there ... the gateposts, I mean?"

He laughed. "Of course. I expect the bodies have been removed by now though. Come on, we're nearly at the top. Take my hand and I'll pull you up."

She had had her head down, watching the path for rabbit holes, and had not seen the ridge dwindle before their advance. Now she was standing higher than she had ever stood before, on top of the world, the soil soft and springy under her feet and the Quantocks, those magic hills that beckoned every day from the churchyard at home, stretching away, fold after fold, into the distance.

She shaded her eyes against the glare. "Is that the sea?"

"It's Bridgwater Bay, part of the Bristol Channel. I suppose you could call it the sea."

"I've never seen the sea."

"You've never –? I don't believe you," Geoffrey said.

"Whatever sort of upbringing are you girls giving the child?" Gabriel said. "A naval officer's daughter and she's never seen the sea — you should be ashamed of yourselves."

"Of course she's seen the sea," Frances said. "She saw it every day when we lived in Plymouth."

"I don't remember Plymouth."

"Didn't the Sunday School go to Minehead last year?"

Frances laughed. "Oh yes. Not Sarah, though. She got so excited beforehand that she made herself sick. She couldn't go."

"Oh dear!" Gabriel said and grinned at Sarah. "Poor old mouse. What a disaster!"

Sarah scowled. "It wasn't just me being sick. Frances nearly was, too. It's always the same. Julia and Annie have to do all the clearing up. Frances can't. She's hopeless. She can't bear to look at a dead chicken even."

"Is that true?" Antony said, prancing backwards in front of Frances. Sarah could tell by the gleam in his eye that he was thinking of practical jokes, planning ways in which he might make Frances faint.

"You disappoint me, Frances," Gabriel said in mock dismay. "All these years I've nourished a secret hope that one day you might nurse me back to health if I were dying and now I learn that you couldn't even smooth my fevered brow."

"Oh, I might manage the fevered brow bit," Frances said. "Nothing more though."

She spoke lightly but Sarah could tell from the angle of her head and the way she walked that she was angry. See if I care, Sarah thought. Serves her right. She shouldn't have told Gabriel about the outing. She's mean. She trailed behind as the others set off along the path, and kicked at the turf. The day was spoilt, ruined by Frances. Now Gabriel would think she was stupid ... a baby ... wish that he hadn't insisted on her coming.

Voices floated back across the ridge.

"I think it's outrageous. It's not as though you lived far

from the coast. I tell you what – why don't we take her on to the sea now that we've come so far? There's a splendid beach at the end of the Quantocks."

"But Gabriel, it's miles," Julia said. "We'd never get back before dark."

"Do we have to? If we found a barn for the night we could ... I often slept in barns when we were campaigning for the Minority Report. Farmers don't mind as long as you ask them first."

They stopped, stood round, looked to Frances as the eldest Purcell.

"Come on, Frances ..."

"Please ..."

"It'd be a tremendous lark."

Frances hesitated. "What would Annie think when we didn't come back?"

"We'll send a telegram."

"It's too far for Sarah."

"We can catch the train back to Dunkery tomorrow if she's tired. Come on, Frances, where's your sense of adventure? Everyone else wants to go. Look at Sarah's face."

"It's her I'm worried about. She's never done much walking. If it were just you and me ..."

Gabriel's mouth curved up, "... you wouldn't hesitate. You are an extraordinary girl. The two of us – can't you imagine the furore that would cause in the village?"

"It would be fun, wouldn't it?" She smiled. "Oh, all right. Let's go. You'll have to make peace with Annie when we get back though."

"You mean – we're going to stay out all night?" Sarah said incredulously. She looked at the distant sea and swallowed. Could she really walk that far?

"We'd better keep some food for this evening," Julia said. "Thank goodness Annie always thinks she's catering for the five thousand. Why don't we picnic here? It's a bit early but we couldn't have a better view."

Sarah nibbled at her pastry. Excitement and the heat had taken away her appetite; she was astonished by the amount the boys ate.

"I must say, Annie does make the most splendid picnics," Antony said. "One of these days I'm going to suggest that Cook goes over to Hillcrest and takes a lesson or two from Annie."

"You'd better be careful," Geoffrey said. "She'll take offence and hand in her notice and then Mother'll be on the warpath. Cook's all right. She's better at dinner parties than Annie."

"As we don't give dinner parties except for you lot . . ."

Gabriel sat down beside Sarah. "You ought to know where we're going, mouse. Let me show you."

His fair hair fell forward as he bent over the map, revealing the pale skin of his forehead where it had been shielded from the sun. White laughter lines fanned out from the corners of his eyes. Between his parted lips Sarah glimpsed the chipped front tooth. He looked up at her and smiled. "You haven't heard a word I've said."

"Yes, I have. What happened to your tooth?"

"My . . .? Oh, that. A cricket ball. At school, years ago."

She wanted to keep him sitting beside her, talking. "Show me where we are again."

Julia and Lucy packed up the picnic things. Gwen lay back on the grass, her hat tipped forward over her face. "Isn't it nice," she said, voice muffled by the straw, "to think that there's no need to rush? We can stay here all afternoon if we want. Do you think Willis will see to the greenhouses if I'm not there?"

"You and the garden," Geoffrey said. "And Frances is as bad. Do you ever stop drawing, Frances?"

"It's the Slade training, I'm afraid," Frances said. "They were very keen on draughtsmanship. They made you draw, draw, draw. It was a very good thing. That's one reason why I think Julia should go."

"Don't let's go into all that again," Julia said. "We can't afford it for one thing."

"Of course we can. We did for me. Besides, you can always pick up the odd prize or scholarship."

"You could, you mean. I'm not in your class at all, Frances. I probably wouldn't even be accepted."

"Don't be silly. You're much better than the average. You'd get in easily with the portfolio you've already got. Can't you persuade her, Geoffrey? It's such a waste if she doesn't go."

Geoffrey's eyes were shut. "I'm not going to try to change her mind. It's her life; let her do what she wants."

"Anyway, I'd be terrified at the Slade," Julia said. "If Frances spent all her time there in tears . . ."

Geoffrey sat up. "Oh come on. Frances in tears? I don't believe it."

"It's true," Frances said. "Everyone was, all the time. The girls, that is. The men just ground their teeth with fury. It was Professor Tonks's fault. He'd stand by your easel, enormously tall with a face like a corpse, and *peer*. 'What's this?' he'd say. 'What's this? An insect?' When you'd put your soul into the beastly drawing! It was even worse if he ignored you, of course. Crying wasn't much good then. You just wanted to curl up and die."

"It wasn't only the Slade though, was it Frances?" Gabriel said, putting away his compass and folding up the map. "Who hated leaving Hillcrest? I can remember you crying into your coffee at Lockhart's you were so homesick."

Underneath the crest of the ridge the stubby bushes spread down towards the bracken below, covering the slope in a prickly carpet. The berries were dull blue, refreshingly sweet. Sarah pressed out their juice with her tongue, feeling the pips gritty against the back of her teeth. If Frances hated leaving Hillcrest for a term at the Slade, could she bear to leave it for ever?

Dust had changed the colour of her shoes, the berries that of her fingers. She slipped slowly down from bush to bush. The sun was very warm; she felt prickles of moisture on the bridge of her nose, gathering under the waistband of her skirt. The voices faded. It was almost too hot, and a long time since she had left her bed.

High above her, its body like an arrowhead against the sky, a lark sang.

Chapter Six

Voices called her out of a dream. Faces circled above her head, black in the sunlight, laughing.

"It's whortleberry juice."

"Sarah, your face!"

"She's turned into an ancient Brit," Antony said.

"Oh, mouse," said Gabriel, pulling her onto his lap. He took out a handkerchief – "Lick that" – and scrubbed her face with it, frowning against the glare, eyebrows meeting over the bridge of his nose. "I hope you haven't made yourself sick."

"I didn't eat very many," Sarah said, wincing at the pressure of his fingers.

"You certainly managed to spread them around. There." He looked at her critically. "I suppose it's an improvement."

"You've ruined your handkerchief," Frances said. "You'd better let me have it – I'll wash it when we get home."

They dusted Sarah down, straightened her skirt, ("I can't think what Annie will say when she sees the mess you're in," Frances said, to which Antony retorted, "That shows terrible lack of imagination, Frances. I know only too well.") and set off towards the sea. Sarah clutched Julia's hand, blinking, not yet properly awake, stumbling over the tussocks. Gabriel strode along the ridge above them, silhouetted against the sky, his hair, gilded by the sun, outlining the shape of his head. He looked like a god, Sarah thought dizzily. A Greek god. Or Achilles marching against the wicked men of Troy.

> *"In Xanadu did Kubla Khan*
> *A stately pleasure-dome decree:*
> *Where Alph, the sacred river ..."*

Once started Antony never knew when to stop. Sometimes he would spend an entire summer afternoon chanting. Latin, Greek, English, poems unknown by poets unheard of – the meaning hardly mattered to Sarah, lying among the tall grasses and giant cow parsley of the Hillcrest orchard, drowning in the torrent of words.

> *"So twice five miles of fertile ground*
> *With walls and towers were girdled round"*

The others had joined in. The rhythm of the words made walking easy. Julia looked down at Sarah and smiled. "All right?"

> *"But O, that deep romantic chasm which slanted*
> *Down the green hill athwart a cedarn cover!"*

"He was thinking of Aisholt when he wrote that," Gabriel said. "He wanted to live in Aisholt but Mrs. C. wouldn't agree."

"Mrs. C.?" Sarah asked timidly.

"Mrs. Coleridge. Coleridge lived for a year in Nether Stowey, did you know? He wanted to be near the Wordsworths who rented a house at Alfoxton, just over the other side from here. It's in a beautiful combe, Alfoxton House. I say, why don't we . . . ?"

"Gabriel!" they shouted.

"It would complete her education, wouldn't it, that and the sea. Do you really think it's too far? Oh well, I suppose you're right. We'll have to do it some other time – next year, perhaps. Why don't we make this walk an annual event?"

"You'd better wait and see what Mother and Father say when we get home," Lucy said.

"Did you know that *The Ancient Mariner* was first thought of here on the Quantocks, mouse? Coleridge was walking over the Quantocks to Watchet with William and Dorothy when the three of them discussed and planned it."

Sarah stopped and gazed at Gabriel in astonishment. "Walking? Worthsworth? Coleridge?" She had never thought of writers as people before, not real people who walked and talked, admired the scenery, had sisters and lived in houses still standing.

She looked down on the patchwork fields and orchards in the valley below, the smoke rising from the villages, the tawny-purple Brendan Hills and Exmoor beyond. One hundred and twenty years ago Coleridge, William and Dorothy must have looked out on the same scene, when the Ancient Mariner and Kubla Khan were nothing more than words floating in the air between them. They, too, must have felt the sun on their faces, the springiness of the turf under their feet, smelt the heather.

"Extraordinary, when you think about it," Gabriel was saying, "the influence the *Lyrical Ballads* had, and yet they were only published because the three of them wanted to go to Germany."

"Rather knocks your ideas on the head, Frances," Antony said. "What about art for art's sake and all that?"

"They'd have written them anyway," Frances said with scorn. "It was only the publishing they did for the money."

> *"Sometimes a-dropping from the sky*
> *I heard the skylark sing."*

Had Coleridge been lying among the bracken and whortle-berries when he thought of that?

"I expect that's how Coleridge saw him," Antony said. "The man from Porlock, I mean. 'Beware! Beware!' He knew he was no good, you see, not to Coleridge anyway. 'His flashing eyes, his floating hair!' That's what laudanum does to you. Poor man from Porlock. He was probably quite ordinary really."

Had she really imagined that words appeared on the page without effort, without conscious thought? Surely not. Yet it was strange to think that other people talked over and helped each other with suggestions and advice as she and Antony did with her essays. It made it seem much more real, knowing that the Ancient Mariner, the Wedding Guest, and the deathly crew, had not appeared in instant inspiration but had only slowly taken shape out of the mist of the imagination.

They came down to tree level for tea. Sarah helped Antony gather twigs and sticks for the fire. The bracken-covered slopes of the combe rose steeply up in front of them as they picnicked in the shade. Sarah was very thirsty but ate little.

"You'll fade away," Geoffrey said. "Still, I suppose that's no bad thing if we're going to have to carry you. How much further do you intend to go tonight, Gabriel?"

"As far as we can before dark," Gabriel said. "I like walking in the evening light. So does Frances."

They sat side by side, he and Frances. His hand lay over hers on the grass, much larger, covering hers entirely. He saw Sarah's glance.

"Such big eyes you have, mouse," he said and smiled. His mouth curved up higher on one side than the other, giving the impression of quiet detached amusement. "What deep thought goes on behind that solemn little face of yours?"

She flushed and looked away.

"I don't know why you bring a map, Gabriel," Julia said. "You know every sheep track and gorse bush, it seems to me."

"I remember the first time I came up here," Gabriel said. He was lying down now, gazing up at the trees, his hand still over that of Frances. "I can't have been more than eight or nine. Father and I came up in the afternoon and stayed till it was dark to see the fires for Queen Victoria's Jubilee. It was an incredible sight – bonfires stretching for miles. I've never forgotten it. I've been coming up ever since. There's something about the Quantocks. I don't know what it is – the Lake District is much more impressive, for instance, but I don't like it nearly as much. I think there's a kind of peace here."

They set off at last, almost reluctantly. Sarah was beginning to falter. Her knees ached. Her feet felt too big for her shoes. Antony led the way.

> *"The western wave was all aflame,*
> *The day was wellnigh done!*
> *Almost upon the western wave*
> *Rested the broad, bright Sun;"*

In spite of the tea, her mouth was too dry to join in. The others drew away, waited for her to catch up, drew away once more. She had difficulty keeping her eyes open.

They were waiting again, Geoffrey carrying Gabriel's rucksack as well as his own. Arms lifted her up.

"All right?" Gabriel said. "Try not to strangle me, there's a good girl."

She had become a giant, riding above them all, looking down on their laughing faces, on a world diminished by her own height. Down in the valley, the trees, woods and buildings threw long shadows across the fields but up here the warm light of evening enveloped the landscape in a golden glow. A faint breeze blew in from the sea, cooling her hot cheeks, playing about her neck, lifting the hair on her forehead.

Gabriel's hair, pale gold, gently rose and fell as he walked, touching Sarah's cheek when she leant forward. The sensation of it, like silk against her skin, made her tremble.

"I think we should look for a barn," Gabriel said. "She's surprisingly heavy. If I remember rightly, there's one near the bottom here."

After the sunlit ridge the barn was dark, full of strange shapes and shadows. The straw scratched and tickled.

"Come and lie down, Sarah," Julia said. She and Lucy had flattened the straw, made a place to sleep.

"Where's Frances?"

"She went to find the farm with Gabriel. Come on. It's very late for you."

"I'm not ready for bed yet," Sarah said, but to please them lay down and shut her eyes.

The stream outside tinkled faintly. Voices murmured by the door. Hens nearby scratched in the straw and clucked. Their beaks pecked on the barn floor, sounding hard and brittle. Antony started on *The Ancient Mariner* for the third time that day. He had just come to the moon—

> *"Her beams bemock'd the sultry main*
> *Like April hoar-frost spread"*

when a voice said, "Do you want a drink, mouse?"

She opened her eyes. Gabriel crouched beside her holding out a mug. The milk was soft against her throat, warm and easy to swallow. "You're not cold, are you, mouse? You can have my jacket if you want." He bent forward and kissed her. "Sleep well."

"I'm not sleepy," she said.

Lucy came over and sat beside her. "Shall I sing you a lullaby?"

"I think I'm a bit old for lullabies," Sarah said and was immediately sorry. She hadn't meant to sound rude or ungrateful. She tried to explain how she had always wondered what people talked about in the evening and how at last she could find out. Tiredness muddled the words, so she stopped, said, "Thank you all the same," and shut her eyes.

And in the end, the day's excitement, weariness, the effect of wind and fresh air, all combined to rob her of her opportunity. Long before the Wedding Guest had departed the words had merged into one word, all conversation swallowed up by darkness, and she slept.

The crowing of the cock woke her next morning, loud and triumphant, outside the barn.

Gabriel muttered, "Do something about that bird, someone."

Somebody groaned, straw rustled, heaved, subsided. Silence.

Sarah sat up. At the far end of the barn was a rectangle of grey. She slid quietly down the straw onto the hard floor and crept towards the door.

It was grey outside. Dark, light, every possible shade, nothing but grey. As she watched, the shadows faded imperceptibly into one tone, like a wash of paint over paper. The sky grew pale, tinged with pink above the outline of the wood. As the world grew lighter the blurred shapes of the trees became separate, absorbed light until they took on colour, became distinct, showed each their own branches, twigs, leaves.

She went back to the barn at last. There was no sound nor sign of movement. She left the others still sleeping and set off down the lane to look for the sea. She had been too tired the previous evening to pay much attention to her surroundings; now she was uncertain which direction to take. She would have to wait for a break in the hedge to get her bearings. It seemed a long way down the lane before she came to a gate; when she did she saw at once that she could not possibly reach the sea before the others woke.

She climbed the gate and sat on the top bar. The clouds' shadows raced each other across the fields to the coast. Once upon a time she had lived by the sea, had looked on it every day, probably paddled in it. Yet she could remember nothing. Where did memories go? she wondered. Perhaps they remained in one's head, locked away in the back of one's mind, hidden. In 1920 would she have forgotten all about today, be unable to recall what it felt like to sit perched on the hard, splintery bar of the gate, gazing across field and sea to the shining strip of land on the horizon? It's not possible, she told herself; I shan't forget, I won't forget. I shall remember yesterday and today until the day I die.

The grass of the field in front of her was dotted with white. Flowers, she thought at first, before she realised that the meadow was carpeted in mushrooms. She climbed down and began picking. They were pink-gilled, glistening with dew. She went deeper into the field, her foot prints leaving dark patches in the grass behind her. She heaped the mushrooms by the lane; when she returned for the third time Julia was on the other side of the gate watching her.

"Hallo," Sarah said. "Can we have mushrooms for breakfast?"

"I don't see why not," Julia said. "We're having breakfast at a farm. Frances and Gabriel arranged it last night. You'll have to ask the farmer's wife to cook them for you. Do you think you've got enough? We ought to get back – Frances thinks you've drowned yourself in the village pond."

Frances rushed out when they came within sight of the barn, seized Sarah's shoulder and shook her so violently that the mushrooms spilled out of Sarah's skirt and fell onto the ground. "You wicked child! Where have you been? You scared us to death."

"Calm down, Frances," Gabriel said. "The child's not come to any harm. Don't look so frightened, mouse. What have you got there – mushrooms?"

"I thought we could have them for breakfast," Sarah said, eyeing Frances nervously. Had she really thought –?

"What a splendid idea! Listen to me, though, mouse. You shouldn't have gone off on your own like that. Suppose you'd got lost. Why didn't you say you were going?"

"You were asleep."

"You could have woken me up."

"You'd have said I couldn't go."

"Oh no, I wouldn't. I'd have come with you."

She stared at him dumbly. To have gone mushrooming, alone with Gabriel...

"Never mind, now," he said. "Don't do it again, that's all. You mustn't worry Frances like that. Come on, Geoffrey, let's see if we can borrow a razor. The others can follow; Frances knows the way."

Frances's anger went as quickly as it had come. She merely sighed at Sarah's skirt, limp, wet and mushroom-stained, and said, "I dread to think what Annie will say when she sees that. Never mind. Go and have a good wash in the stream."

The water was peaty brown as it ran over the stones and so cold that it took Sarah's breath away.

"Oh, Frances," she said as Frances tried to comb the tangles out of her hair, "I watched the sun come up when you were all still asleep. It was wonderful."

Frances smiled. "You're not tired? I thought you'd be exhausted."

"No. I am hungry though. I'm sure I can smell bacon from here."

Mrs. Mackenzie disliked eating out of doors. It was the result of her upbringing: in India one kept out of the sun. Meals at the Rectory were always in the dining-room. Despite her disapproval the Purcells frequently ate on the verandah at Hillcrest, shielded from the sun by the wisteria that climbed up the trelliswork and trailed over the glass roof. But no meal at Hillcrest had ever tasted as delicious as this one. Mushrooms, eggs, thick pink bacon, bread and jam, mugs of sweet tea: there was more than anyone could possibly eat in spite of the hunger that fresh air had given them.

Afterwards they helped carry the plates into the kitchen and then sat unmoving in the sun. A troop of ducks waddled up to clear the crumbs, pushing flat beaks into the grass, until there were none left and they retreated quacking to the stream. Bees moved through the purple-flowered clematis on the farmhouse wall. From the dairy came the regular thump of butter-making.

At last Gabriel yawned and stretched. "I suppose we ought to get going. I thought that after we'd been to the beach we could walk along the cliffs to Watchet and show Sarah the Ancient Mariner's harbour. We'll catch the train to Dunkery from there – it's too far to expect her to walk all the way home. All right? Put your drawing things away, Frances, while I settle up."

The lane took them below the Quantocks ridge, past fields of wheat and stubble. The going was easier than the day before but Sarah's legs felt strangely stiff, almost as if they did not belong to her at all. It was very hot. The farmer's wife had talked of drought. Sarah was glad when they reached the wood and were able to continue in the shade. The path wound through the trees, dropping steeply all the time, and without warning came out into the open. They stood at the foot of the bay.

The cliffs rose up round them, blood-coloured, streaked with slanting lines of blue and green. Pebbles covered the beach, mottled, marbled in alabaster, grey and pink. Beyond the pebbles stretched the sand and beyond the sand, at last, the sea.

Sarah did not know, now, what she had expected. In one of her storybooks at home there was a picture of Grace Darling. Waves, cliff-high, towered over the boat, sending spray up to the top of the page. It was not like that at all. The sea was flat, smooth as a piece of grey paper laid across the Channel to Wales.

"It's a marvellous place for fossils," Antony said. "Come on, Gwen."

"I suppose you want to paddle," Frances said to Sarah. "Take off your shoes and stockings and we'll tuck your skirt into your knickers. Do try not to get too wet." She looked up at Gabriel, eyes laughing. "Look at that sea. Doesn't it make you want to . . . ?"

He grinned. "You behave yourself, my girl. I'm sure your delightful Slade friends would throw off every stitch they had on and dive headlong into the sea but you're not going to embarrass me like that. Not here. Not in public."

"It's not public. It's only us."

"Even so. Why don't you take off the child's skirt? Then at least she'd have something dry to put on if she got wet. Come on, mouse, let me help." He squatted down beside her, his head level with hers. "You'd think they could make it less complicated for a child," he said as he fumbled with the fastenings. "All right? Step out of it then."

She felt shy and awkward standing half-dressed on the beach beside Gabriel. But Gabriel was not looking at her. He was staring at Frances and she at him as if they had never seen each other before, as if they could not believe what they saw in each other's eyes.

"Can I paddle now?" Sarah asked. No-one answered. After a moment she left them and walked slowly towards the sea. A light breeze played round her bare legs. The sand was soft and wet. When she wriggled her toes sand came up between them like small brown worms.

The sea shushed up the beach and washed gently over her feet, ripples taking on the colour of the sand beneath, silvered round curved edges. When she stopped, the water took the sand from under her feet and ran back with it down the beach.

She walked further out, the sunlit villages on the distant coast beckoning her on. The sun, and the reflection of it off the water, was extraordinarily bright, giving the world an unearthly clarity that filled her with sensations never before experienced. It was as though time had stopped and caught her up, contained her, in eternity.

Then it was gone. The world clicked back into place. She was a child again. The sea lapped round her ankles, seagulls cried in the sky. For a moment she felt bereft, lost. She wanted to describe, share with someone, that brief moment of ecstasy, but where could she find the words with which to do it?

Frances and Gabriel came slowly across the beach towards her, holding hands. Frances had let down her hair: caught by the wind it blew across her face towards Gabriel. Sarah saw the expression on their faces as they drew near and knew then that there was no need for words.

The train rattled under the edge of the Quantock hills on its way to Taunton from the sea. Sarah leant against Gabriel and shut her eyes. The tweed of his jacket was rough against her hot cheeks, smelling of seaweed and salt. She drifted into sleep and drifted out again. Conversation ebbed and flowed round her like the sea, loud words, soft words, words without meaning. When she opened her eyes she could see Frances sitting opposite. Frances's head rocked gently with the movement of the train. Her windswept hair, escaping from pins carelessly put in, fanned out over the upholstery, rich chestnut against the green. Frances smiled across at Gabriel, her eyes still holding the memory of that moment on the beach:

> *"For she on honey-dew hath fed,*
> *And drunk the milk of Paradise."*

But Mrs. Mackenzie had no knowledge of honey-dew, or Kubla Khan either. She greeted Purcells and Mackenzies with tight lips and cold eyes on their return to Huish Priory.

"I'm sorry if you were expecting lunch," she said. "The table has already been cleared. I can't have the servants held up in their duties by your lack of punctuality. And as for last night ... I think you had better wait in the Rector's study. I'll talk to you there. No, not you, Sarah. You weren't to blame. Bertha, take the child over to Annie, please. At once."

"On the warpath proper, she is," Bertha told Annie in the Hillcrest kitchen, but Annie gave no answering smile.

"I'm not surprised. I might have known they'd finish up doing something silly. Wearing Miss Sarah out like that – they haven't got the sense they were born with, none of them. Well, I'm having none of it. Upstairs you go, miss, this minute and put yourself to bed."

Sarah gaped. "*Bed*? But, Annie, it's the afternoon still."

"Bed's what I said and bed's what I meant. I'm not having you whining for days to come, all because they've worn you out."

"I don't whine," Sarah said indignantly, but she knew better than to argue with Annie.

She sat at her window, looking across at the Rectory, waiting for the others. For two days she had been grown-up, one of them. Now she wasn't even old enough to be scolded. It was my fault, she thought, I was the one to blame. It was all because I couldn't remember what the sea was like. She shut her eyes. I don't care how angry they are, Mrs. Mackenzie or Annie. Or even Mr. Mackenzie. It was worth it. I'd do it again.

Chapter Seven

The scratching of Antony's pen nib sounded surprisingly loud in the study where he and Sarah sat having their lessons.

Beyond the bare winter trees outside the Rectory window, the green-grey slate roofs of Hillcrest crouched on the brow of the hill like an animal about to pounce. It gave Sarah a comfortable feeling, looking out at Hillcrest from her table in the study. It was as if she had two homes, Hillcrest and the Rectory, held together by the ribbon of the Rectory drive that twisted and turned between the trees and shrubs on its way to the road and the brow of the hill.

It was a pleasant room, the study, furnished with heavy, practical furniture and shelves of leather-bound books that sagged as they climbed the walls from floor to ceiling. It was Sarah's favourite room; in it she felt safe, cosseted, loved.

The only wall without bookshelves was covered with photographs, group photographs of schoolboys, under-graduates, theological students, the Bishop's staff. There was one photograph, recording the triumphant conclusion to a tiger hunt, that particularly fascinated Sarah. When she was small and new to lessons at the Rectory she used to go over to that photograph as soon as she arrived, to gaze at the tiger. He was dead. He didn't look dead as he lay under the Governor-

General's foot, staring at the camera with a toothy grin, but he was. Stone cold dead. That's what the wolves are, she told herself every morning. Quite dead. It was easy to say in the daylight of the Rectory study; less easy to believe when she was alone in her bedroom at night and Annie and her sisters too far away to hear her if she called.

The tiger photograph was the only suggestion of India in the study. It was Mrs. Mackenzie who came from India, whose family had lived and worked there since the time of Plassey. Mr. Mackenzie had merely been visiting the sub-continent when he fell in love with the Governor-General's daughter who had been told to entertain her father's guests.

"Good at her job even then," Antony said. "Trust the White Queen. You can understand what Father saw in her, of course, but why do you think she married him?"

"I don't know," Sarah said, thinking wistfully of Gabriel, whom Geoffrey said all the girls adored.

"Well, they lived happily ever after," Antony said. "Though if it hadn't been for the climate disagreeing with Father I'm sure she'd rather have stayed in India. She says it's so damp in Somerset. I'd have thought the monsoon might be a bit damp too, myself." He sighed. "It's a funny thing, marriage. I don't know that I'll bother. When you think of girls like Helen of Troy it doesn't seem worth the trouble, does it?"

Everything came back to marriage these days. Sarah's thoughts hovered round the subject like a tongue playing with a loose tooth. What was it like being married, she wondered, being in love? What did you do? What did you find to talk about all the time?

She knew about Bill Roberts and Mary from Jess, but their leisurely courtship seemed inappropriate for Frances and Gabriel. When she tentatively broached the subject with Antony he said, "Frances and Gabriel, do you mean? Well, she paints and he reads Horace aloud to her. Don't look at me like that – it's true. I came across them last summer. Frances couldn't stop blushing when she saw me and Gabriel was furious."

She didn't believe him. There must be more to it than that. The idea was ludicrous. Frances didn't understand Latin for one thing. What did frighten Sarah considerably, when she thought about the conversation later, was the fact that although she had phrased the question in the most general terms Antony had immediately assumed she was referring to Frances and Gabriel. She remembered their faces that day on the beach. Did other people know what she only guessed at?

The scratching from Antony's pen stopped. Antony was listening, as Sarah was, to the sounds beyond the study door: Cook's heavy breathing as she emerged from the basement on her way to the morning-room to discuss the day's menus with Mrs. Mackenzie. Bertha's footsteps climbing the stairs to turn out the bedrooms, Hilda brushing down Mr. Mackenzie's coat.

"Half an hour, Hilda." Mr. Mackenzie said the same thing every morning before he left to take prayers and the scripture lesson at the school. A gust of cold air blew into the house. The front door closed.

Antony laid down his pen and grinned across at Sarah. "Now we can relax."

There was still their work to do, of course: Mr. Mackenzie never went out without making sure they were properly occupied. But the rest of the household was busy with the morning chores of Rectory life and paid little attention to the chatter of the two children in the study.

Sarah avoided mentioning the dread subject these days. She knew that Antony, secure in the comfortable permanence of his own world, would never understand the fears and insecurities that the mere possibility of Frances's marrying produced in Sarah.

Annie said that Antony was spoilt because he was clever and supposed to be delicate. "Delicate!" she snorted. "He's stronger than the rest of that family put together." She had a soft spot for him all the same and encouraged Sarah to go off with him in the afternoons. "One intense person in the family's quite enough, thank you very much. At least he gets your nose out of your books and there's nothing wrong with being a bit of a tomboy now and then."

Which was why, when Sarah tore her skirt on the big oak beyond the spinney, it was to Annie that she showed it, to Annie she confessed.

"Next time you go climbing trees, miss," Annie said when she had cobbled the pieces together, "kindly take your skirt off. No more sense than you were born with, that's your trouble. What do you think knickers are for? And if you're in the spinney just make sure Nanny Mackenzie doesn't see you. She'd have a fit and tell Mrs. Mackenzie into the bargain."

It was just as well Annie couldn't see Antony and Sarah pretending to be Jason and Tiphys on the topmost branches, swaying perilously in the wind as they scanned the horizon for the first sight of the towers of King Aeetes' palace, or she might well have been the one having the fit.

Cook came out of the morning-room and padded down the stone stairs to the kitchen. Mr. Mackenzie's footsteps sounded on the doorstep. But instead of coming into the study he was waylaid by Mrs. Mackenzie.

"I'd like to have a word."

"Can it not wait? The children . . ."

"It won't take a moment. It's about Gabriel's letter. Henry, I'm worried."

Antony raised his eyebrows at Sarah as Mr. Mackenzie's footsteps sounded across the hall into the morning-room.

"Well?"

"I think you should write to Gabriel and warn him."

"Warn him? About what?"

"He was in London last week. For several days, he says. You can guess why, can't you? She was up there too. They must be meeting."

Surprise sounded in Mr. Mackenzie's voice. "You're not suggesting they shouldn't – that we should keep them apart? My dear! How could we object? He saw her often in London when she was at the Slade."

"That was different. She was a student then, not much more than a child. We wanted him to keep an eye on her. It's not like that now, you know it isn't. She's – what, twenty? Self-willed, independent. Dangerous."

"That sounds very melodramatic," the Rector said. "I think that you are letting your feelings run away with you."

"You don't believe that Gabriel's interested in her ...?"

Mr. Mackenzie said slowly, "I wouldn't go so far ... They obviously enjoy each other's company. I doubt whether it's anything more than that."

"Then we should make sure that that's how it stays. You must write and let him know what we think. Ask him to be sensible. Tell him not to see too much of her."

The Rector's voice was gentle. "My dear, Gabriel is a grown man. He has his own life to lead. We can't interfere."

"Sometimes it's one's duty to interfere. Are you suggesting that we should sit back and watch him make a terrible mistake? Don't you realise what an unsuitable match it would be, a girl like that? What sort of don's wife would she make?"

"An intelligent one, I would have thought."

"What about politics – he's always had leanings in that direction. Suppose he goes into politics?"

"Knowing his views one hopes he won't. My dear, you yourself have always said that Frances is very capable. If she sets her mind to do something, she does it and does it well. I'm quite sure that she could cope with politicians. For all we know, she may well have been involved with Gabriel in the Fabians."

"You'd be happy then, to have her as a daughter-in-law?"

"I didn't say that. It would certainly not be a marriage like ... ours, for instance. That may not be what Gabriel wants. After all, they have grown up together. He has had every opportunity to discover her faults as well as her virtues. If you consider the matter calmly I'm sure you'll realise that Gabriel is not the sort of person to rush blindly into marriage."

Mrs. Mackenzie hesitated. "I still think that you should write."

"It may have the wrong effect. If the idea is not already in his head, it would certainly put it there. My dear, we must be careful. An ill-spoken word, a hasty letter, could have quite the opposite effect to the one you intend. Gabriel has always been amenable to reason, even as a child, but occasionally he can be very obstinate. Do you not remember that time he had set his heart on a full-sized cricket bat? He was only eight or nine – we thought him too young."

Mrs. Mackenzie sounded shocked. "A cricket bat is hardly the same thing as a wife."

"Of course not. Though to an eight-year-old ... What I'm trying to say is that if Gabriel is really determined ... He always got his own way in the end, even over the cricket bat. We don't want to turn him away from us."

There was a moment's silence.

"You're her guardian," Mrs. Mackenzie said. "You could refuse permission."

"She'll be twenty-one next year. I can't stop her then. Besides – has Gabriel mentioned marriage?"

"Not as far as I know. I want to make sure that he never does. Oh Henry, I'm fond of the girl, you know I am, in spite of her being so difficult. She can be charming, she's attractive, capable. But she's ... different. She doesn't see things the way other people do. Sometimes I think she's possessed. I want the best for my sons. I warn you, I'm not having her as a wife for Gabriel. When all's said and done, she's nothing but a bohemian." Mrs. Mackenzie's voice rose. "A bohemian. Admit it. Is that what you want for your son?"

The front door bell rang. Hilda's footsteps sounded across the hall. The morning-room door closed.

Antony got up and shut the study door. "Well, well, well. I didn't know the White Queen had it in her. She did go up to London last week, didn't she?"

"She?"

"Frances."

Sarah kept her eyes on her book. The words merged into a grey block on the page. "Yes."

"Did she see Gabriel?"

"I don't know. She went to see some of her Slade friends."

Antony laughed. "That's what she said."

Sarah flushed. "Frances doesn't tell lies. If she said that, it's true."

"All right, all right. There's no need to bite my head off. You're right, of course. I'm sorry. Still, she might've wanted to see Gabriel, too, mightn't she?"

"I suppose so." She said, despairingly, "Oh, I wish . . ."

"What?"

"Why can't Frances be nice to your mother? She does it on purpose, I'm sure she does. Makes your mother angry, I mean. Gwen and Julia don't do it."

"It's just one of those things. Don't get so upset. Mother's as bad – always bossing Frances about when she knows perfectly well Frances won't stand for it. I must say, I do like Frances – you never know when the sparks'll fly with her around. Oh, come on, Sarah, you don't really mind, do you? I think it's a laugh. Tell you what, let's have a bet on it. Who would you back in a fight – Gabriel or the White Queen?"

Mr. Mackenzie came in before Sarah could reply. "Not at your desk, Antony?" he said, gently. "Have you finished the work I set you?"

Antony looked sheepish. "Well . . ."

"Let me see how much you have done."

Antony grinned at Sarah as Mr. Mackenzie sat down at the

desk. He didn't care. It didn't matter to him what his parents thought of Frances. A joke, that was all it was to Antony.

Sarah gazed down at her book with unseeing eyes. Did Mr. Mackenzie think Frances was a bohemian too? A bohemian.

I can't ask, she told herself, I mustn't let them know I know, but somehow I must find out what that word means.

Chapter Eight

Every year Christmas beckoned like a bright light through the grey mists of autumn. The prospect this year was almost too dazzling to contemplate, for this year Sarah was to be allowed to dine at the Rectory with her sisters.

"I don't know, I'm sure," Annie said when Sarah danced into the kitchen to tell her the news. "Staying up till all hours. You're only ten after all. Miss Frances was fifteen before she was allowed to stay up for dinner. Spoilt, that's what you are."

"Oh Annie, don't be mean. Gabriel says I always get left out. Antony's stayed up at Christmas for years and years and years."

"Master Antony gets away with murder. That's no reason why you should do so. You'll be cross-patchy for days afterwards, I shouldn't wonder."

"I won't, I promise I won't. I'll be so good you won't believe I'm me." She said, placatingly, "Besides, I shan't be very late. Mrs. Mackenzie said I couldn't stay all evening, I'd have to come home after dinner."

"Humph," was all Annie would say to that.

The awful question occurred to Sarah as soon as the first excitement had worn off, and worried her for days.

"What shall I wear?" she asked Frances at last, and Frances said, as if it were a matter of no importance, "Oh, we'll find you something, I expect."

She was teasing of course but Sarah was not quite sure until Christmas morning when she picked up Frances's present to her and prodded it with anticipating fingers. "Oh Frances, it isn't, is it? Is it ... a dress?"

Frances laughed at her. "You didn't think we'd let you go in your old holland, did you? Aren't you going to see what it's like?"

"I know it'll be splendid. A new dress!" She pulled off the ribbon, tore the wrapping paper, held it up ...

Perhaps that was when Christmas started to go wrong.

She had known from the beginning what she wanted: a liberty dress in silk, low at the neck, low at the waist, with pleats and pin tucks; the sort of dress Lucy might have worn when she was ten. But this ...

"Try it on," Frances said. "I want to see if the hem is straight. It's only tacked. I'll sew it properly this afternoon."

"It's much too long," Sarah said, trying to keep her voice steady.

Julia gave her a quick glance. It would never occur to Frances that the dress might not be perfect but Julia would notice at once. "We thought you'd like it," she said. "It's grown-up. You're always wanting to be grown-up."

It was high round the neck, the sleeves full, caught into a cuff at the wrist. The smooth cashmere material was gathered into a high waistline. When she sat down it would fall in soft folds. Attractive to paint. That was how Frances looked at clothes, from the point of view of paint. She was only interested in the material, the way it hung, how it fell, the effect of light on its texture. After Christmas she would make Sarah pose.

Sarah looked down at Frances as she knelt on the floor checking the hem. I don't want to be painted. I don't care about pictures. I want clothes that I like, clothes for me.

Matins on Christmas Day was Sarah's favourite service but this morning the magic had gone. Perhaps Gabriel's behaviour had something to do with that. No winks, no smiles, not a glance in Sarah's direction. Sarah was hurt as well as puzzled. She was about to whisper to Frances when she realised that Frances was keeping her own gaze averted from the choir stalls. The sermon, normally a time for smiling across at Frances or Sarah, Gabriel spent staring down at his hands. Not even during the interminable prayers for peace in Ireland did he look towards the Purcells' pew.

After the service the congregation grouped outside the porch. In no mood to exchange greetings, Sarah kept her eyes firmly fixed on the vestry door but before the choir could emerge Frances took her hand, muttered something about helping Annie and started along the path for home.

"Do wait, Frances," Sarah said. "I haven't wished Gabriel a happy Christmas yet."

"You'll see him tonight."

"Christmas'll be nearly over by then. Please Frances, I want to see Gabriel."

But Frances refused to wait. Pink-cheeked and determined, she dragged Sarah away. When Sarah looked back from the gate Gabriel was talking to Miss Tuck. His back was to Hillcrest . . . he had not even noticed Sarah's absence.

Nothing escaped Annie's eyes. Today, because it was Christmas perhaps, she waited until evening when she was alone with Sarah helping her get ready for dinner at the Rectory before she spoke.

"You know what I'm going to say, Miss Sarah," she said, brushing Sarah's hair with long, hard strokes. "Ungrateful, that's what you are. Don't think the good Lord hasn't noticed, because He has. There's a black mark waiting for you up in heaven. You're lucky to get a new dress at all, my girl, let alone a smart one like that. There's no reason why you shouldn't have had one of Miss Gwen's that she's grown out of. We're not made of money in this house, you know. It's make do and mend and I've never heard your sisters complain. And Miss Frances went to a dint of trouble, finding the right material and making it up when she could have been working."

"I'm not ungrateful, Annie. It's just that it's ... well, it's not fashionable."

Annie tipped the ribbons out of Sarah's ribbon-box onto the dressing-table. "Hark at Miss Hoity-Toity. Who are you to be fashionable, eh? It's the sort of thing Miss Frances's London friends wear when they come and stay, isn't it?" She selected a ribbon that went with the dress and tied it in a bow round Sarah's hair, saying more gently, "You may not like it, love, but it suits you very well. The colour brings out the lights in your hair."

Sarah surveyed herself in the mottled surface of the mirror. "Does it really? Do you think Gabriel will think I look nice?"

"H'm. Mr. Gabriel's only got eyes for Miss Frances these days."

"He hasn't. He didn't look at her once this morning, not even during the prayers."

"And what were your eyes doing wandering round church when they should've been tight shut? Eh?" But she was smiling. "Off with you now, love. Perfect manners, mind, all evening."

Frances was already in the hall in a dress of soft white wool that followed the shape of her body and fell in soft folds to the floor; unfashionably plain and yet somehow stunning. Her hair shone like the skin of a newly opened chestnut and when she moved her head the emerald ear-rings that had been Mother's winked in the light.

"Oh Frances," Sarah said, "if only I looked like you."

"Be thankful that your teeth don't stick out like mine!"

Julia and Gwen came downstairs. Annie fussed round them, checking that they had their shoes, seeing hems were straight, buttons done up. Sarah's heart knocked against her rib-cage.

"Can't we go yet?"

"The church lights only went out a few minutes ago. We must give them time to change."

Two baskets sat on the hall chest, Christmas presents for the Rectory servants in one, presents for the Mackenzies in the other.

"Let's go," Gwen said. "It doesn't matter if we're early. Mrs. Mackenzie and Lucy won't have gone to evening service."

It was very dark outside. Banks of black clouds massed behind the trees. High in the sky one or two stars glimmered faintly. I wish there were a moon, Sarah thought. A moon would be romantic. Frances walked silently beside her, her face pale above the blackness of her cloak. She looked no different from Gwen, from Julia. Why had Mrs. Mackenzie called her bohemian?

Sarah touched Frances's arm. "You will be nice to Mrs. Mackenzie tonight, won't you?"

Frances smiled down at her, wide-eyed, innocent. "I don't know what you mean."

"Yes, you do. Please, Frances. Promise."

"I'm always nice."

The fir tree that Sarah and Antony had helped bring in from the garden stood in the Rectory hall, stretching up beyond the first floor landing. Its decorations moved in the warmth of the candles and glittered in their light. Underneath the branches, in glorious, exciting disarray, the presents waited to be opened.

"Let me take the baskets," Hilda said, the superior expression of head parlourmaid gone in the excitement of Christmas.

"Fancy Miss Sarah!" Bertha said. "You'd never believe how time flies. Looking so pretty, too."

"And very grown-up," Geoffrey said.

He looked grown-up himself, quite changed since the summer. Perhaps it was the clothes that made a difference.

Sarah had never seen him in evening dress before. He seemed a different being from the unhappy, fearful boy in the fruitcage. Cambridge had been better than he anticipated, Sarah knew, for he had told Julia so. "He sounds surprised," Julia said when she read his letter, sounding surprised herself. Yet his hands as he helped Julia remove her cloak still seemed too big, and somehow out of control.

Evening dress made no difference to Antony. He slid down the banisters, tie askew. "Heavens," he said when he saw Sarah, "there's a stranger in our midst. 'Is this the face that launched a thousand ships?' Which reminds me ..." He bent down and whispered in her ear, "I'll show you where Gabriel's hidden the mistletoe after dinner. Don't say a word to Father, mind. He doesn't approve of mistletoe – pagan rites and all that."

She giggled, shivering with excitement. None of the men could take their eyes off Frances, she noticed, wishing that Gabriel would turn round and look at her. He kissed Frances, murmuring something that made her blush, greeted Gwen and Julia and at last came over to Sarah, looking her up and down with a crooked grin. "Well, well, well."

She was taken into dinner by Mr. Mackenzie, her arm stretched up to rest on his. She had never seen the dining-room at night before. The beauty of it, silver and glass sparkling in the candlelight, took her breath away.

"Come on, Sarah," Lucy said. "Gabriel's arranged the seating this year. You're in the place of honour by Father. Frances, you're opposite, next to Gabriel."

"I see so little of you these days," Mrs. Mackenzie said. "Sit by me tonight, Gabriel. Geoffrey won't mind changing places."

Sarah feared that she might faint with excitement, knew that she would never be able to eat. She must remember everything, every little detail.

"I don't think Sarah ought to be having wine ..."

"Oh, come on, Frances," Gabriel said. "A little won't hurt the child. She can't drink to the King with an empty glass."

They drank to absent friends before they began the meal, remaining quiet and thoughtful for several minutes. Most of the Mackenzies' relations were far away in India. All my friends are here in this room, Sarah thought. No, not all. There's Jess. And Annie.

"I think we should drink to our youngest guest and welcome her among us at last," Mr. Mackenzie said. "Sarah. May she have a great future."

They turned towards her, raised their glasses. She was confused, not sure what to say, hot-cheeked.

"What sort of future do you think she's likely to have?" Geoffrey asked.

"Oh, she'll undoubtedly go on to university," Mr. Mackenzie said. "Now, are we all served?"

Sarah stared at him. "University? Cambridge, do you mean? With Gabriel and Geoffrey?"

"Oxford, I think. It has rather more to offer a woman, and by the time you get there they may well grant women degrees as well as men. I think Oxford would suit you better than Cambridge."

"Dear me," Gabriel said mockingly. "What a family! Three artists and a bluestocking. You'll all be wanting the vote next."

She was dazed. Their faces turned towards her, Gabriel smiling beyond the candelabra, Frances unsurprised, amused. University? Surely not. "Do you mean it? You're not teasing?"

"My dear child, do I 'tease'? Of course, it's early to say yet, but you certainly have the potential. Provided you work hard over the next few years I see no reason why you shouldn't gain a place."

"It's something to aim for, isn't it, mouse? You can do all sorts of things with a degree. You'll have to start thinking about a career."

"I don't know how you can say things like that, Gabriel," Frances said. "You are so hypocritical."

Her voice was quiet, conversational. Sarah had no idea that

[83]

anything was wrong until Mrs. Mackenzie laid down her knife and fork with great deliberation and said in steely tones:

"May one enquire what *that* is supposed to mean?"

Antony nudged Sarah. "Here we go again."

Frances flushed. "Gabriel knows exactly what I mean."

"Gabriel may know: I do not. I am tired, Frances, of your constant criticism . . ."

"I'm not criticising, Mrs. Mackenzie, I'm saying what I think. Gabriel says one thing when he means another. I consider that hypocritical. Perhaps you don't."

Gabriel's face was flushed, his eyes angry. "If only you'd listen. You latch on to the first thing I say and jump to all sorts of ridiculous conclusions. You're so obsessed with your own ideas you won't listen to anyone else at all. What makes you so sure you're always right?"

"Because I am right. You make a great song and dance about careers for women, you go on about studying and the importance of training and all that sort of thing. You agree with everything I say, but when it comes down to it the only career you think worthwhile is marriage."

"Well, isn't it?"

She said angrily, "Of course it isn't."

"Frances . . ." Julia said.

"Don't interrupt."

Geoffrey said nervously, "More wine anyone?" and was ignored.

"What sort of state would the world be in if women didn't get married?"

"I'm not talking about ordinary women. I'm talking about women with a career. How can they give that up? What sort of career is marriage – running round after a man?"

"I think you're becoming over-emotional, Frances," Mr. Mackenzie said. "Marriage is a splendid career. Look at Mrs. Mackenzie. Where would this parish be without her to run the Mothers' Union, the G.F.S, the Sunday School? Who would help the children find work when they leave school?"

Frances hesitated. "That's different."

"Why is it different?" Gabriel asked. "You're confounded by arguments again, aren't you? Women are so illogical. I keep on telling you, nowadays you can have it all ways."

Frances said with scorn, "Oh yes, I know which way you want it."

"I have no idea what started this ... discussion," Mr. Mackenzie interrupted, "but I do think that any disagreement between the two of you should be kept apart and not brought to the dinner-table, particularly on a festive occasion such as tonight. And you, Frances, have obviously forgotten the part Gabriel played in your own career. Mrs. Mackenzie and I were very against your going to London. It was Gabriel who persuaded us that we would be wrong not to encourage your talent. You owe him a great deal."

Frances flushed. "Yes," she said after a moment. "I do know that. I'm sorry."

But she sounded unrepentant and refused to look at Mr. Mackenzie, staring with bright eyes at the wallpaper somewhere over Sarah's shoulder.

At the sideboard Bertha tried, unsuccessfully, to control her giggles. Even Hilda was smiling. Why did Frances ... why couldn't she be nice? Sarah gazed miserably down at her plate. You promised. You said you'd be nice. You did promise.

Mrs. Mackenzie's hand shook as she signalled for the vegetables. Bertha went round the silent table, mouth pursed, eyes towards the floor; no-one wanted more.

Sarah played with her food, appetite gone.

"Leave it," Frances said in a low voice. "It doesn't matter."

Sarah looked at her sister across the table. At that moment she hated her; hated her bitterly.

Chapter Nine

The pungent smell of pine needles and hot candle grease mingled with traces of turkey and plum pudding in the hall. Discarded Christmas paper and unwanted ribbon hid the rich patterns of the Indian rugs.

One present, flat, rectangular, remained under the Christmas tree. It was not difficult, when all the other presents had been exchanged, unwrapped, exclaimed over and put aside, to work out that this must be from Gabriel, a joint present for all the Purcells.

Sarah knew what Annie would say and didn't care. If only, if only Gabriel could have given her something for herself alone. A pencil, an exercise book (but only Antony knew she wrote), something, no matter how small or cheap, that she could hold in her hand and know that he had chosen specially for her.

He picked up the parcel and held it for a moment, balancing it on his hand, before holding it out to Frances. "I want you to open it."

She laughed up at him, anger apparently forgotten, chin tilted in the air. "I can guess what it is."

"Of course. I always said you were more intelligent than you looked. But only the *sort* of thing, not what it is exactly."

"Do hurry up," Julia said. "I want to see . . ."

Frances knelt on the floor, the material of her dress falling in folds like rippled water round her. Her fingers fumbled with the ribbons, tore the paper, pulled away the wadding.

"Well?"

"Oh, Gabriel." She sounded as if she were about to cry. "Oh, Gabriel."

It's her present, Sarah thought. It's not for us at all. He bought it for her.

"I couldn't wait to give it to you. I thought Christmas would never come."

"But Gabriel . . . a Matisse . . ."

"I knew you'd like it." He knelt on the floor beside her, watching her face. "You do like it, don't you?"

"Oh, Gabriel . . ."

It was as if no-one else in the hall existed, as if the quarrel had never happened.

Mrs. Mackenzie cleared her throat. "May we see . . .?"

Frances came out of her trance. "Yes, of course." She leant the canvas against the brass pot that held the Christmas tree.

It was a still life: three oranges and a bunch of grapes in a wooden bowl on a covered table, with a coffee-pot in the background.

"Oh, Gabriel," Gwen and Julia exclaimed together. Just like Frances. Sarah was silent. A picture, a stupid painting, when Hillcrest was crammed with such things.

Mrs. Mackenzie frowned. "But Frances — it's a child's daub. You can't admire that sort of thing, surely? It's not even well drawn."

Frances shook her head. "That doesn't matter. It is well drawn, as a matter of fact, but not clearly defined. It's the light that matters, not the subject. Look at the tablecloth, the way the whiteness of it throws the fruit into the foreground, makes it live. They're really there, those oranges. You could pick one out of the bowl."

"Well ... perhaps," Mrs. Mackenzie said. "But the actual drawing ..."

"You can always take a photograph if that's what you want."

Mrs. Mackenzie shook her head. "I'm afraid I'm old-fashioned. I like a picture to be a picture."

"Never mind, Mother," Gabriel said. "Ninety-nine per cent of the population of this country would agree with you."

"So would you, three years ago," Frances said. "You thought that that exhibition was rubbish until I took you along the second time."

"We all know what a Philistine I was in those days. I've learnt better since you took my education in hand."

"Or worse," Antony said, "depending on which way you look at it. Do let's go into the drawing-room. I want to hear *Ragtime*. Did you know Gabriel bought the music the other day?"

In the drawing-room the fire blazed in the grate. Sarah sat on the hard settee blinking sleepily at the flames, trying to keep her eyes open, while the others gathered round the piano. The notes tinkled out in an odd, catchy rhythm, unlike anything Sarah had ever heard.

> "*Oh, ma honey, oh, ma honey,*
> *Better hurry and let's meander,*
> *Ain't you goin', Ain't you goin'*
> *To the leader man ...*"

"Do we have to listen to such a dreadful noise? And the grammar ... "

"Oh, Mother," Antony said. "You're being old-fashioned again. It's Gabriel's favourite show so it must be all right."

"That is no recommendation," Mrs. Mackenzie said. "I used to think that Gabriel had good taste, but I don't know what has happened to him recently."

"I'm surprised at you, Mrs. Mackenzie. How can you say such a thing?"

Aware of a sudden hush, Sarah opened her eyes. Frances and Mrs. Mackenzie faced each other in front of the drawing-room fire, Frances impudent, mocking; Mrs. Mackenzie, if she could ever be said to glare, glaring.

"Gabriel still has perfect taste," Frances said, "especially in people. I'm sure you'll agree that his taste in people is impeccable."

Mr. Mackenzie said quickly, "It's not the most suitable music for Christmas Day, Lucy. Find something more appropriate, my dear. Frances, I understand that Gabriel took you to see *Boris Godounov*. Tell me what you thought of Chaliapin."

Gabriel came over to Sarah.

"I'm sorry about the present, mouse. It wasn't the best choice for someone who doesn't paint, was it? Or have you decided to be an artist like your sisters?"

"I'm no good at painting. Frances says it's very strange when there's so much talent in the rest of the family."

His eyebrows went up. "Is that so? She can be very crushing sometimes, can't she? I shouldn't take it to heart if I were you. University will be much more fun, don't you think?"

"Yes," she said, but doubtfully. What exactly did you do, at university?

"You do like the painting, don't you?"

Frances liked it; wasn't that what mattered? Yet he seemed to want her approval and it was quite a nice picture really. She nodded. "Yes. And I see what Frances means. It is alive, isn't it, not like that stag over the mantelpiece. You can't imagine him even breathing."

He laughed. "Quite so. But don't say that in Mother's hearing."

In the warm glow of mutual understanding Sarah thought, Gabriel wouldn't mind my asking, and said quickly before she could change her mind, "Gabriel, what's a bohemian?"

"Bohemian?" He hesitated. "Doesn't Father tell you to look up words you don't know?"

"The dictionary said it was a native of Bohemia."

"Well? You know where that is, don't you? Anything else?"

"A gypsy."

"Yes?"

"Someone who leads an uncon ... unconven ..."

"Unconventional?"

"Yes. A – what you said – life."

"That seems fair enough."

She was puzzled. "Is that all?"

"No. A bohemian can be a poet, or writer; an artist of some kind. Someone whose work is more important to him than food or warmth or comfort. Puccini wrote an opera about four bohemians starving in a garret – I'll take you to see it when you're older, if you like." He frowned. "Why do you want to know about bohemians, mouse?"

She avoided his eyes. Across the room Frances perched on a stool, laughing and talking with Mr. Mackenzie. Like the Matisse painting, she was vivid, alive. A bohemian. Sarah could feel Gabriel watching her.

He said casually, "Some people might even call Frances bohemian."

It was as if he could read her mind. "Would you?"

He was silent for a moment. "Yes, I think I would. Her work is all she cares about." He smiled. "If Father hadn't insisted on her having decent digs in London she would probably have starved in a garret, don't you think?"

She felt as though a great load had been lifted off her mind. She was suddenly light-hearted, marvellously happy. "Doesn't it matter then, being bohemian?"

"Of course not. Well, to some people, perhaps. Listen, mouse. People aren't the same, you know. They have different ideas, different likes and dislikes. Take Fergus Donne at the Manor. He's only interested in things with four legs or wings that you can ride or shoot. He won't think much of bohemians – or intellectuals either, for that matter. Whereas my mother" – he was watching Sarah closely – "approves of intellectuals.

Of course she does: she thinks I'm one of them. But because she doesn't understand modern artists – bohemians, we'll call them – she neither likes nor approves of them. It doesn't mean that there's anything wrong with them. I like bohemians myself and intellectuals, too. Not to mention little girls who ask odd questions when they're as pretty as you are tonight, mouse."

She blushed, confused.

Lucy clapped her hands. "Time for charades."

"I think Sarah ..." Mrs. Mackenzie began.

"Let her stay, Mother," Lucy said. "It is Christmas. Besides, it'll make the numbers even for once. Come on, Sarah, you can be on my side. Frances, you'd better go with Gabriel, otherwise he won't play. I'll have Antony to help Sarah and –"

"I'll have Julia," Gabriel said. "Really, Lucy, you might let someone else organise the teams occasionally. Take your lot out, I'm much too comfortable to move." He smiled at Sarah. "All right now?"

It was dark in the hall. The smell of wax hung in the air although the candles on the tree had been blown out long ago. Draughts drifted down the stairs and hovered above the rugs, rustling forgotten scraps of wrapping paper. Sarah sat on the bottom stair and wrapped her skirt round her ankles. She yawned and shut her eyes as she leant against the banisters. Why couldn't the others agree? What was wrong with 'matrimony'?

"They might not like it," Gwen said.

"I think it's silly, Antony. A bit juvenile, don't you think?"

"Not at all. I think you're a lily-livered lot. It's only a joke, after all. They won't mind. And Sarah can be the bride – that dress couldn't be better. You'd like to be a bride, wouldn't you, Sarah?"

"Yes, please," she said at once.

"There you are then. Take her upstairs, Lucy. You can find a bit of veiling or something that'll do, can't you?"

She sat in front of Lucy's dressing-table, watching Lucy pull out drawers and open wardrobe doors. A fire was laid in the grate. Did Bertha light it every evening? The room was shiveringly cold. In the triple mirror the room was reflected three times over, six candles guttering, three beds turned down, three triangles of white sheet, three pillows, soft, inviting. It must be very late, hours after her bedtime. Perhaps she should have gone home after all.

"Here we are," Lucy said. "Why don't we pin up your hair? Whoever saw a bride with hair down her back?" She pulled out the ribbon, twisted the hair in her hands and piled it on top of Sarah's head, fixing it in place with pins and a piece of veiling.

"There," she said, stepping back. "The bride. How do you like yourself?"

Sarah's head felt heavy, awkwardly balanced on her neck, unsafe. She stared at the image in the mirror. A dream, a promise, a stranger. Shall I look like you when I'm grown-up? she asked that other, unknown self. The stranger smiled uncertainly back.

"I don't think it's a good idea," Lucy said when she and Sarah came down to the hall. "She looks too like . . ."

"It's too late for second thoughts," Antony said. "Do come on. You've been ages. We've done the first syllables already."

The evening had become a dream, blurred at the edges, and she herself a ghost, the spirit of Christmas yet to come. Firelight flickered in front of her tired eyes, voices murmured from a great distance. She moved where Antony told her to move, said what he told her to say and thought of the stranger in the mirror upstairs. Pretty, Gabriel had called her. Pretty. Did he think her pretty now? Beautiful? She searched for his face among the shadowy spectators.

He had turned away from the charade and was sitting very still, watching Frances: Frances, upright, rigid, eyebrows a black bar across her white face.

Sarah faltered, stopped, burst into tears.

Frances's voice was like ice. "The child's over-tired. She should be in bed."

Confusion. Chairs pushed back, dresses rustling, people whispering.

Gabriel appeared in front of her, holding out a handkerchief. "I'll take her back to Annie. Blow your nose, mouse, and we'll go home. It must be long past your bedtime."

Night had transformed the churchyard into an unknown world of shadows and mysterious rustlings. Who knew what evil spirits lurked behind the bushes that stretched out their spidery branches to snatch at passers-by? The black bat, night...

Gabriel strode ahead, silent with his own thoughts. Sarah tripped in the tussocky grass as she ran after him, trying to keep up. The hem of her dress was wet and clammy round her ankles.

The notes of the song tinkled in her head:

> *"Come on along, come on along,*
> *Let me take you by the hand"*

At the end of the churchyard the chestnut trees stood like a ghostly army of skeletons against the banks of cloud. They shook their arms at Sarah and sighed. "Come on along, come on along ..."

The elms in Tinker's Lane hung threateningly over the front garden. An owl hooted in the copse beyond the stables. The sound of the door knocker took a long time to die away on the night air. It was impossible to see Gabriel's expression in the darkness of the porch. What was he thinking? What would he say?

"Here we are, Annie. One little girl safely delivered. Very tired, I'm afraid."

Annie was ready for bed, her hair tied in a single plait. "I'm not surprised, at this hour. Come on in, love. Say thank you to Mr. Gabriel for bringing you home."

Sarah clutched at Gabriel's sleeve. "I'm sorry ... I didn't mean ..." But she was not sure what she didn't mean.

"I know. It was Antony's doing, the little fool. And it's a pity Frances is so touchy." He patted Sarah's head. "Don't give it another thought, mouse. Sleep well. And you, Annie."

"What was that all about?" Annie asked as she helped Sarah undress.

"I don't know." Sarah's voice trembled. Her eyes filled with hot tears. "Oh, Annie, Frances was so cross ..."

"Come now, love," Annie said. "We don't want tears on Christmas Day. Though it's been a funny old Christmas when all's said and done. Not what some people hoped, that's certain."

Sarah watched her hang up the new dress. "Do you like Gabriel, Annie?"

"Do I – that's a silly question, miss. Of course I do."

"Frances hates him."

"That's not true, Miss Sarah, and you know it."

Sarah remembered Frances's eyes across the dinner-table. "Sometimes she does."

"Ah well. Up and down, that's one of the troubles of Miss Frances's age. When you've grown up a bit, I daresay you'll understand. Now lie down and shut your eyes. I'll let you sleep on in the morning."

Sarah stared into the darkness and thought about Frances. She had not known Frances could be so silently, so frighteningly angry. For no reason – that was the worst of it. She crawled down the bed and made a nest for herself under the blankets. Her breath was warm and damp on her cheeks. She cried herself to sleep.

She woke, struggling to breathe. A voice said, "It's all right," and suddenly her head was out in the cold. There was a shivering yellow glow of light. She thought that she was still asleep, that the Madonna over the mantelpiece had come alive and was standing by her bed.

It was Frances, her hair a copper halo in the candlelight. "Are you all right?" she said gently. "Annie said ..."

Sarah gazed up at her, her eyelids heavy with sleep, trying to remember something back in the dream.

"What's that round your neck?"

Frances's hand went to her throat. "It's a pendant."

"I haven't seen it before, have I?"

"Gabriel gave it to me for Christmas."

"But the picture ...?"

"That was for all of us. I think he wanted me to have something of my own."

She smiled down at Sarah. Her face was soft, luminous, full of warmth and love. The picture of it remained in Sarah's dreams long after she had closed her eyes and drifted back to sleep, long after Frances herself had gone to bed. It was as though the Madonna had stepped out of the picture frame over the mantelpiece and come down to comfort her and wipe the tears away.

1914

Chapter Ten

They made a gloomy group in the middle of the bank holiday festivity: Mr. Tasker, the retired barrister from Clay Court, donning his courtroom face for the occasion, Colonel Sherwood, drooping, if a man of such military bearing could be said to droop, the Rector sober-faced and Gabriel unnaturally solemn. Even Sir James from the Manor, usually the most cheerful of men, looked grim and despondent.

Sarah, hovering round them hoping to attract Gabriel's attention, thought that they would never stop talking and such dismal, dreary talk, too.

"A catastrophe . . ."

"There's still time," Mr. Tasker said, but not hopefully.

"I think not," said Gabriel. "Once Russia mobilised . . . I'm sure it's inevitable now."

"Young David Hancock from Northridge called at the Rectory last night to say good-bye," the Rector said. "He's been recalled to Devonport."

"We can see the Minehead line from Clay Court, you know," Mr. Tasker said. "Two trains ran into Taunton yesterday. Full, as far as we could see. They must be bringing the territorials back early from summer camp—on a Sunday, too."

"Weren't you in Germany earlier this year?" Colonel

Sherwood asked Gabriel. "How were things then?"

"It was taken for granted. A matter of time, they all thought, and assumed that I did, too."

"I don't like it," Colonel Sherwood said. "It was all very well in my day. Now – airships, dreadnoughts – it's a very different kettle of fish."

Lady Donne swept into the group with a rustle of silk. "Not still talking your gloomy talk of war! You are supposed to be enjoying yourselves this afternoon. I can't think why you are so *serious*! What have we got an army for? Or the navy? Leave war to the people concerned."

"You don't understand, my dear," Sir James said. "This time we shall all be concerned." He sighed. "Never did like the Germans to tell you the truth. Always thought the Kaiser an untrustworthy sort of fellow."

Sarah tugged at Gabriel's sleeve. "What's Germany got to do with Ireland?"

He frowned down at her. "Ireland?"

"Isn't there going to be civil war?"

"What? No, no, it's nothing to do with Ireland. Here – get yourself some strawberries and cream." He pulled a shilling from his pocket. Seeing Jessica hovering in the background he said, "Buy some for Jessica, too."

War was forgotten as the two girls made their way past the village band beating out the syncopated rhythms of *Ragtime*, past the coconut shies, the needlework stall and flower exhibition, towards the marquee under the trees.

Lady Donne appeared in front of them, like a vision from another world, a world only glimpsed at in the pages of society magazines that occasionally graced the Rectory drawing-room. Unlike Sir James, who spent the greater part of the year at the Manor, Lady Donne disliked the West Country and as soon as she had done her duty – opening fêtes, distributing prizes to schoolchildren or half crowns to villagers, entertaining the local gentry – she returned with relief to the pleasures of the London season.

She was doing her duty now when she said, "I am pleased to see you are maintaining your standard in needlework, Jessica."

"Yes, ma'am."

Lady Donne waited. "Well?"

"Thank you, ma'am."

"We mustn't forget our manners, must we, Jessica?"

"No, ma'am."

"She was waiting for you to curtsey," Sarah whispered as Lady Donne, forehead creased in disapproval, moved away.

"'Tis silly, curtseying," Jess said, her mouth set in an obstinate line. "*You* don't curtsey to her so why should I?"

"Frances says we need only curtsey to royalty."

Jess giggled. "Lady Donne thinks she'm royalty all right. What's that Master Antony calls her?"

"The Red Queen. You know, 'Off with his head!' Only I shouldn't have told you that, Jess. Promise you won't tell anyone. Or about Mrs. Mackenzie being the White Queen."

"Good thing her's in London such a lot. It'd be like having two queen bees in the village, else. One of 'em'd have to go."

Lady Donne might be making a royal progress through the meadow but there was no doubt who was the queen bee today. The bank holiday fête was always in aid of the church and it was Mrs. Mackenzie who organised and ran it. She was in the refreshment tent now preparing for the rush, seeing that the Manor kitchenmaids kept the trestle tables replenished with clean crockery, checking that the ladies of the parish were behind their tea urns. Not as grand as Lady Donne perhaps – and today she was a working woman – but elegant, immaculate despite the heat and as fresh as though she had just arrived on duty. Only those, like the elder Purcells, who had been on duty since the early morning would know that Mrs. Mackenzie had been on her feet all day.

She saw Sarah at the strawberry table and came swiftly over. "I thought that you had already had yours, Sarah."

"Gabriel told us to come now," Sarah said, clutching her

plate. It wasn't a lie, and why shouldn't she have more if she paid for them?

"Take them outside then," Mrs. Mackenzie said, more kindly. "We're too busy in here for children."

They found a place to sit hidden from Mrs. Mackenzie's sharp eyes, between the guy ropes at the back of the tent. The grass was still wet from the morning rain; Sarah felt the damp seeping up through her skirt. It made her feel pleasantly naughty: she knew what Annie would say.

"Gabriel says there's going to be a war," she said and licked her bowl with her tongue. She might as well be really wicked while she was about it.

"What sort of war?"

"Civil war in Ireland, I thought, but he says not."

"Oh, Ireland. All them prayers Rector goes on with, every Sunday – I get proper sick of Ireland."

"Well, it's not Ireland this time. Germany, I think. Or Russia, did he say?"

"I've got a cousin's a soldier," Jess said, "lives over Dunkery way. He's in South Africa now."

The complicated relationships of Jess's family had always amazed Sarah. There seemed to be Hancock cousins of one kind or another in every part of the world. "He'll be coming back soon, I expect. Have you finished? I want to win a coconut."

Despite Antony's tuition, Sarah was not much good with a ball. The coconuts remained untouched until Gabriel arrived and knocked one down for her at his first attempt.

"Any time," he said when she tried to thank him, clutching the prickly prize in triumph. "Do you know where Frances is?"

"She was helping with the teas the last time I saw her," Sarah said. "Can I come with you?"

Outside the refreshment tent the band, scorning the weaker stuff on offer inside, had provided themselves with flagons of cider. They lounged on the grass, faces flushed, collars unbuttoned, navy and gold caps thrown on the grass. The instruments laid out beside them glinted in the sun. Gabriel stopped.

"All set for Saturday, Bill?"

Bill Roberts stood up, grinning. "Aye, thank you, sir."

"The ringers sound in good form. I couldn't hear myself speak last night."

"It'll be a good do. We're putting a guard round the tower come Friday to make sure of it."

"A guard?"

"Didn't no-one tell you about our Susan's do last year? Dad took the bell ropes away, night afore the wedding. Didn't think young Chedzoy good enough, he said. So there weren't no church bells, see? They Chedzoys have long memories. I'm not having them spoil my wedding."

Gabriel laughed.

Bill said, "Could you spare a minute, Mr. Gabriel?" He came away from the band and lowered his voice. "Do you reckon the army'll need farriers? With this business overseas, like."

"I imagine it'll need everyone it can get. There's talk of sending the yeomanry overseas, I believe. Are you –?" He glanced over towards Mary Hancock who was sitting near the band. "I'd wait a week or two before doing anything, if I were you. You don't want to rush into anything at the moment."

"Aye. Reckon she might take on a bit. Thanks, Mr. Gabriel. I'll think about it."

Frances was behind a trestle table in the refreshment tent, Geoffrey and Julia keeping her company. She waved across at Sarah. "I've kept you some strawberries and cream. Do ask your mother if I can go, Gabriel, I've been here for hours."

Gabriel came back as Sarah was scraping her plate with her spoon, lacking the courage with Frances watching to use her tongue. "Well, well," he said. "You are in favour. Performed nobly, she said ..."

"I enjoyed it. I'm exhausted now, though. Have you ever noticed the way the light changes when it comes through the canvas? Reflected light – I must remember next year. I could paint in here in the morning when there weren't too many people about."

"Lord, Frances," Gabriel said. "Don't you ever think of anything else?"

Sarah thought of Frances when she was in the swingboat with Jess. Would Frances be able to snatch a painting as she went up and down? How splendid it would be to have a picture of what she could see now: the countryside laid out beneath her like a bird's eye view of west Somerset. The churchyard chestnuts hid Hillcrest from sight. Trees shielded the Manor too, except for brief glimpses of the gravel drive, the striped grass of the lawn and a peacock strutting across it. The trees dipped down to the village where smoke hung upright in the sky above hidden cottages. Beyond the village, fields and woods stretched across the valley to the Quantocks on the horizon, magic hills glowing in the evening light.

Up towards the sky she went and down again, hair lifting, the air cool against her neck. So God must see the world, His people in miniature, scurrying like ants on the ground below, clearing up the debris of the fête, folding cloths, taking down stalls, carrying trays of dirty crockery into the Rectory. She caught sight of Frances and Gabriel disappearing into the jungle of Rectory rhododendrons.

The band had abandoned *Ragtime* long ago. Now it returned to its first love:

> *"Some talk of Alexander and some of Hercules,*
> *Of Hector and Lysander..."*

"Let's have another go," Jess said when she and Sarah staggered out of the boat. Her hair was blown across her flushed face, lying against her mouth.

"Not now," Sarah said, the strawberries heavy in her stomach. "I'm going to find the others."

The refreshment tent was empty. It glowed inside with a dull yellow light; was warm and stuffy with the collected heat of the afternoon. Fragments of sandwiches, crumbs of cake had been trodden into the flattened grass. In one corner, dismantled trestle tables waited for Sir James's men to carry them back to the Manor. There was a vague, secretive feeling contained under the canvas roof, of regret for one more day past, another church fête over. Sarah stood in the middle of the marquee, breathing in the atmosphere, alone and quiet.

When she went outside the shadows lay across the meadow. They reached out towards the Rectory trees and the church, and crawled imperceptibly up the pink walls of the tower until only the flagpole remained shining like gold against the colourless sky.

Annie appeared at Sarah's side. "Time for bed."

"Can't I stay till it's dark? Jess hasn't gone home yet."

"That's as may be. And I'd call it dark now. You've got a proper peaky look to you – over-tired I shouldn't wonder. Or too many strawberries. Come along, love, don't spoil a good day."

Daylight was fading fast. The gloomy figure of Shattock, the Rectory gardener, ambled round the edge of the meadow dipping his taper into the potted meat jars that were strung between the trees. One by one the candles flared up until they made a golden necklace of fairy lights suspended in the night air.

The band, thirst and hunger slaked in the Rectory kitchen, tramped back up the drive. The time for dancing had arrived. Under the hedge by the gate Antony and Gwen practised a tentative waltz, waving as Sarah went reluctantly by.

Tom raised his arms, said, "Ladies and *Gennulmen* – take your partners for the first waltz" and brought the drumsticks down on the parchment with a bang that sent the rooks screaming up from the trees.

The village girls in their pale-coloured dresses drifted like a chorus of ghosts over the black grass. Frances and Gabriel merged into one...

The drumbeat came into the lane, rolling round the hollow

space between the trees, thudding inside Sarah's ribcage. Even in her bedroom she could feel its vibrations. She rested her face against the cool glass of the window. Glimmers of fairy lights winked at her through the trees.

The music changed. She knew all the tunes; if the wind was in the right direction heard them faintly during band practices, knew the words from evenings spent round the Rectory piano.

> "I hear the mission bells above...
> Ringing out our song of love..."

The trumpet notes swooped up and up, before dropping down like water, note on note quickly falling until they drowned in the deeper music of the bass-horns. She pictured Bill Roberts, face flushed with cider and song, trumpet glinting in the half light, drunk with the power of his own music, while out on the grass Frances and Gabriel moved in each other's arms.

> "Ramona ... I need you, my own."

Chapter Eleven

Now that life was so busy, so full of excitement, Sarah could not imagine how she had managed to keep herself occupied in the days before the war.

There were meetings in the parish room, bandaging sessions at the Rectory, sewing afternoons at the Manor. After the first of these, when Lady Donne, never having held a pair of dressmaking shears in her hands before, cut out twelve left legs because none of those watching had the courage to tell her that it was necessary to turn the pattern over for the right leg, the sewing party confined itself to nightshirts for Taunton Hospital and flannel shirts for recruits. Even Sarah could cope with the seams and hems of those.

There were plenty of other things to do, too, apart from the organised activities. Knitting, for instance – scarves and belts for the West Somerset Yeomanry and socks in scratchy wool that brought your hands up in an ugly red rash.

Sarah walked with Gwen and Julia for miles through dusty country lanes in the hot summer sun, picking blackberries to make blackberry and apple jam for the Belgian refugees. ("Lord love us," said Annie, "let's hope the poor souls are getting more than jam to keep them going or they'll go back where they came from. Which may be no bad thing, when all's said and done.")

The Scouts went off to help patrol the main railway line from London, as a result of which the eldest Shattock boy gave himself airs for weeks. Anyone would have thought he'd held the German army at bay single-handed instead of walking up and down an empty railway line for a few hours in the twilight.

Men enlisted immediately: Gabriel and Geoffrey, of course, and Bill Roberts only days after his wedding. Mr. Podimor left his sister to replace him at the church organ which lengthened services considerably, as she had to have several attempts at each hymn and the psalms almost always defeated her. Mr. Mackenzie was waiting for a suitable occasion to propose, tactfully, that Lucy might do better.

Willis, the Hillcrest gardener, went next, to the consternation of everyone but Gwen.

"I'm not sorry," she said. "He'd never do what I wanted or try out anything new. We're always last in the village with peas and beans, you know. It's a relief, in fact. Now I can do what I want."

"But Gwen," Frances said, "the garden's huge. We can't possibly manage on our own and how can we find somebody else at a time like this?"

"We won't try," Gwen said. "We'll have Willis's wages to play about with. I'll speak to Miss Ross. I expect one or two of the older boys would be pleased to earn a bit of money after school and I'm sure that Sir James won't mind lending us one of his men occasionally to help with the trees in the copse and the heavy pruning. Willis wasn't much good, you know. I don't think he liked taking orders from a woman; I'm sure he only enlisted to get away from us. But you'll all have to help more than you do. You use the garden a lot, Frances, what with your painting and the flowers but you only do a bit of deadheading now and then."

"All right," Frances said. All three of them were taken

aback by this new, masterful Gwen. "You'll have to tell us what to do though."

Rumours ran up and down the village street, details changing at every cottage. Russian soldiers had been seen with snow on their boots in trains from Scotland, Germans were chopping off the hands of French boys to dry up the supply of soldiers, and doing things to Belgian girls that no-one dared mention in front of Sarah. And as for Count Alpbach...

He had lived among them so long, mixing with the gentry, riding with the staghounds, that people forgot Count Alpbach was foreign. After August the fourth they remembered, and remembering realised that his departure for Germany at the end of the previous month had a sinister significance. He must have known war was coming and hurried home with secrets for the Kaiser.

There was a Marconi wireless installation in the cellar of his house, someone said; no, it was in the stables, said someone else. Three hundred rifles were stacked in a bedroom ... thousands of gallons of petrol hidden away...

Sir James and his chauffeur drove over to Farrance House to investigate, arriving as a groom was leading out a string of horses for exercise. Quick thinking and neat driving on Hawkins's part put Sir James's car across the gateway, blocking the horses' exit. When help arrived and the house was searched no wireless installation was found, nor rifles or petrol, nothing to indicate espionage at all. There were the horses, however, twenty thoroughbreds – "splendid beasts" according to Sir James who had frequently hunted with Count Alpbach – which were confiscated from their absent owner and sent over to France to be used against his countrymen.

"It makes me quite ill," Frances said, "all those people who called themselves his friends. And those beautiful horses, dragging guns ..."

"Do you mean to say he wasn't a spy after all?" Sarah said. She was bitterly disappointed.

Geoffrey came home on weekend leave, astonishing Sarah. He appeared to have grown taller, bigger in every way and gained confidence, despite the lack of uniform – weapons too, he said. These days he had even gained control of his hands.

Gabriel at least had uniform when he arrived, though Frances thought little of it. "Such a pity," she said, smiling sideways at Mrs. Mackenzie, "that the uniform should be khaki. I'm afraid khaki really isn't your colour."

Gabriel laughed. "Be thankful it isn't blue serge. It was all Geoffrey's lot had when I last saw him and blue serge doesn't do anything for Geoffrey's looks."

Criticism of Gabriel, however indirect, was the last straw as far as Mrs. Mackenzie was concerned. She sucked in her breath, and pulled back her shoulders.

"I have been patient long enough, Frances, but I cannot remain silent a moment longer. Your behaviour these last few weeks has been outrageous. Have you no patriotism? You stand there and mock – oh yes, I saw you laughing over the pyjama trousers. What have you done to help the war effort, may I ask?"

"Now then, Mother," Gabriel said, easily. "Frances . . ."

"You keep out of this, Gabriel," Frances said. "It's all very well, Mrs. Mackenzie, but what's this war about, can you tell me that? Stupid little boys playing games, that's what it's about. Fooling around with guns, showing off, getting carried away. Well, they've gone too far now – it's too late to pull back now, isn't it? What do you think matters, Mrs. Mackenzie? Trade? Battleships? Power? You don't care if Rheims cathedral's been destroyed, or Kreisler's lost an arm or all the manuscripts of Louvain have gone up in smoke. None of that matters so long as we can beat the beastly Germans. We stand up in church and roll off the ten commandments and then what? Half the men in England are given guns and told to go off and murder the Hun. The world's gone mad."

"You may not agree with what is happening," Mrs. Mackenzie said. Her voice was tightly controlled but her chest heaved, "but in England's hour of need it is your duty to help in whatever way you can. Sewing, bandaging . . ."

"Oh, bandaging – that really is a farce, isn't it? Those poor old dears – do they really think the Germans are only going to hit below the knee?"

"I am quite sure that we shall be able to call on you, with your experience in the life class . . ."

Mrs. Mackenzie stopped.

Frances's face was paper white. After a moment she said, "I'm not staying here a moment longer. Come along, Sarah," and walked very fast out of the Rectory and along the path, muttering, "Stupid woman ... stupid woman ..." through the churchyard until she reached the gate. Her hands shook so much that Sarah had to pull back the latch.

Frances stopped on the topmost step. "I suppose you think I'm wrong. Well, I'm not. I'm right. Perhaps when you're older you'll be able to see. All the same, I did behave badly. I shouldn't have said all that. She makes me so angry, sometimes I think I could kill her."

"I think you're wrong about the khaki," Sarah said; it was that that had hurt most of all. "I don't care anyway, even if it isn't his colour. I think he looks splendid."

The frown went from Frances's face. "Of course he does. He always looks splendid. Oh Sarah, you love him too, don't you?"

Her hair glowed like copper in the sun that came low over the Hillcrest roofs as she danced in a circle on the road, arms outstretched like a being absorbed in some ancient rite. "Oh Gabriel, I do love you so much."

Antony came into the study like a gust of fresh air. "You'll never guess what," he said. "I've enlisted."

The Rector, opening his mouth to reprimand his son for being late, was startled into silence.

"Isn't it splendid?" Antony said. His head was held high, his face flushed – like a conquering hero, Sarah thought.

"They won't take you," Mr. Mackenzie said at last. "The scarlet fever – you're not fit enough."

"I am. They have."

"They *can't* take you – you're under age."

"Yes, well. I'm sorry about that, Father. I had to fib."

"Oh Antony, not you. The others, but not you."

"You don't understand. If I don't hurry, it'll be too late, I'll miss it all. I can't possibly miss it, you must see that."

"What about your studies?"

"The war'll be over by Christmas. What difference will six months make? It's worse for Geoffrey, leaving in the middle of Cambridge."

Mr. Mackenzie's skin was sallow, stretched over his cheekbones. "I can't tell your mother. You'll have to tell her yourself."

"I thought she was going to faint," Antony told Sarah later. "She didn't, of course – White Queens don't faint. It's ridiculous, really. She expected the others to enlist – she'd have been furious if they hadn't. Why such a fuss about me?"

"What did your father mean about the scarlet fever?"

"Oh that. Well, you know what parents are. No, I suppose you don't. They're inclined to fuss, you know. Of course I am the youngest and I did nearly die, but that was years ago. They still tend to think ... Actually, if they'd thought I'd fully recovered they'd have sent me off to school like the others. I would have hated that. Your time's not your own there. As a matter of fact, the old man was quite right. I tried to enlist ages ago, after Mons, and they wouldn't have me. Said my chest measurement wasn't big enough."

"But how . . . ?"

"It had to be thirty-five and a half inches at the beginning. Now it's down to thirty-four and a half. I suppose they want more men. So that made it easier. But the first time I went along, when I was turned down, the recruiting sergeant told me about a gymnasium which would help. I've been borrowing George Cross's bicycle and going in every week. It's made all the difference. If I do the exercises for an hour and then bicycle like the wind into Taunton my chest's all right. It goes down again after a bit, though, so I had to be quick."

"So if you went back now, they'd turn you down?"

"I expect so. Silly, isn't it? As if the size of your chest makes any difference to killing Germans."

She was filled with conflicting emotions: admiration for his determination and the way he had kept his attempts secret despite the close Rectory household, envy because he could go away to fight and she could not, disappointment that he had not confided in her; but most of all she felt excitement, a quivering passionate excitement that sent shivers running up and down her spine at the thought of Antony marching behind the band on his way to war.

Antony's departure had an unexpected, unbelievable consequence. Sarah was to go away to school. She could scarcely believe her ears and gaped incredulously at Frances when she told her the news.

"Mr. Mackenzie thinks that you should be working with other children. It was all right when Antony was here, although he was so much older, but now he's gone you've got nothing to spur you on. That's what Mr. Mackenzie says; he's afraid you might get lazy. It's supposed to be a good school; some girls go on to university every year. Mr. Mackenzie thinks it would suit you, and I'd go by his judgement in things like that."

Although not the boarding-school she had always longed for it was almost as desirable, being fifty miles away, in Bristol. She would live with the aunts ("great-aunts really, they're sisters of Mother's father") and walk to and fro every day.

"It's only a few minutes away, they say. Bristol isn't all that far either, not like London. You'd come home in the holidays, of course, and it might be possible for you to come home for the weekend occasionally too. What do you think? You would like to go, wouldn't you? It seems the most sensible thing to us, but it's up to you."

She mustn't hurt Frances's feelings, mustn't let her see how much she wanted to go, how the mere suggestion of it made her want to dance and sing...

"I don't mind," she said.

Frances eyed her doubtfully. "The headmistress has suggested that you should spend a day there first, to see if you liked it and if they thought you would fit in. There are one or two problems, you see – you're older than they usually take for one thing, and you're ahead in some subjects and behind in others. It'd be easier to sort out if you're there, I gather. We thought you could spend the day in school and stay with the aunts for the night before and after."

"All right," Sarah said. "When can I go?"

"There doesn't seem much point in waiting," Frances said. "How about the beginning of next week?"

Chapter Twelve

The pigeon had lost a claw. Where the foot should have started the leg came to an abrupt end, scales overlapping in an ugly, stubby lump. Sarah thought of Frances, who hated all imperfections and would have quickly turned her head at the sight. What had happened to the claw, she wondered. Severed by a moving train? Trapped under trolley wheels? The pigeon, quite as agile as any of its two-legged companions, hopped about the platform unperturbed by the crowds, the bustle and clamour of Temple Meads, far less nervous than Sarah who had never imagined that a railway station could be so enormous or so noisy. The huge glass roof, arching high above the platform and the tracks, magnified and threw back the racket of manoeuvring trains, escaping steam and general hubbub of people so that her ears were overwhelmed by noise. If Aunt Bessie called, would she ever hear?

At Hillcrest pigeons were enemies – varmints. When you saw one in the garden you rushed at it, waving your arms – a stick if you had one – screaming as loudly as you could. If you were Fergus Donne you shot it. Today the strange surroundings turned the enemies into friends. Sarah watched the one-legged bird as it hopped about on the platform in front of her and wished that she could be as calm, as unimpressed by the grandeur of the station. She tried not to think of all the terrible things that might have happened to Aunt Bessie. Suppose she had been run over by a tram? Or had dropped dead in the street? John the carrier had died on his cart one day and no-one knew until his horse arrived home with the cart and its load behind it. If such a disaster had overtaken Aunt Bessie would anyone remember Sarah waiting, patient but apprehensive, on Temple Meads station?

"Sarah? Sarah Purcell? Oh, my dear child."

She had expected someone with a name like Bessie to be fat, or plump at the very least, but Aunt Bessie was not much taller than Sarah herself, and very much thinner – like a doll, a black-clad, black-bonneted, fragile china doll. Her skin, as she kissed Sarah's cheek, was soft, dry, powdery.

"Isn't this exciting – Herbert's granddaughter! I can scarcely believe it." She held Sarah away from her. "Now then, let's look at you, child . . . Yes, I can see the resemblance, really I can. Poor Laura. How was your journey? Fancy, how brave . . ." Her chin quivered with excitement. She never stopped talking. There was no need for Sarah, suddenly shy, to say anything at all.

"How much luggage have you, dear? Is that all? We don't need to take a cab with that. Come along then – I can't wait to get home and show you off to Maudie."

Sarah found her voice. "Frances said I must let her know I'd arrived safely."

"Yes, of course. We'll send a telegram. I should have thought of that. They'll be anxious, I expect. Think of it – all that way, alone. How old are you? Eleven? Gracious me, I must have been twenty at least before my father allowed me out on my own. Wait here, child, while I see to it."

Sarah thought of home as Aunt Bessie disappeared round the corner into the post office. She pictured Miss Tuck panting up the hill, heedlessly squashing the last chestnuts of autumn under her flat feet, pushing open the Hillcrest gate, marching up to the front door; heard the sound of the doorbell echoing round the house. Frances would be painting as usual in the stables and Gwen would be in the garden, both too far away to hear. Today was baking day – it would be Julia, preparing the dough in the kitchen with Annie, who would be thinking of Sarah, wondering how she was getting on. It would be Julia who would open the telegram ...

Standing outside the station, looking down the incline towards the city, Sarah could almost smell the warm, yeasty fragrance of the kitchen. She felt momentary panic: what was she doing here? Then she remembered. She was going to school at last.

To someone who had never ridden in a tram before the journey back to the aunts' house was almost too exciting: when they had to change at the Tramway Centre, Sarah was terrified that she might become separated from Aunt Bessie and get on the wrong tram.

Aunt Bessie talked all the time, head darting from side to side, black bonnet bobbing. "That's the Colston Hall ... the Prince's Theatre ..."

Sarah smiled, trying to hide her fright as the tram lurched round a sharp bend with a horrible grinding roar.

"There's the drill hall – where the men enlist, you know – and the art gallery. Frances would be interested in that."

Sarah, twisting back to look at the drill hall, missed the art gallery. She was reminded of Antony. What was he doing at this moment? Had he received her letter, did he know that all her dreams were about to come true?

They left the tram near a building that resembled a Grecian temple and continued on foot, in the residential part now, walking past large houses set back from the road. It was very quiet; one or two people walking purposefully along, the occasional horse and trap. Sarah's bag grew heavier every minute. Her arms ached.

"Nearly there," Aunt Bessie encouraged as they turned off the main road. "We're coming to the square."

The square was empty, without life. The terraced houses

were indistinguishable, one from the other, with their black painted front doors and wrought iron balconies standing out from the first floor windows.

Sarah gazed up at the number painted on the glass between the delicate tracery of the fanlight over the front door. Number eight. So many houses grouped round the square — were they all the same inside as they were out, differing only in their number? In Huish Priory no two buildings looked alike, not even the cottages lining the main street.

The maid who opened the door resembled the house: tall, thin, stretched out like a piece of elastic. Her cheeks went in instead of out and her nose was so pinched that Sarah wondered how there could be anything in it but bone. Her expression was fierce, quite unlike the cheerful, friendly expression with which Annie greeted visitors.

"This is Sarah, Violet," Aunt Bessie said. "Laura's little girl. We must give her a warm welcome."

"Of course," Violet said but added no welcoming words of her own and the way in which she surveyed Sarah made Sarah feel that she was even dirtier after the journey than she felt.

Violet turned away from Sarah as if she were of no interest. "I've given Miss Ellison her lunch. It didn't seem sensible to wait any longer. She's having her afternoon sleep now. Please don't disturb her."

"Oh, Violet." Aunt Bessie drooped. Her chin wobbled. "I was so looking forward to introducing Sarah."

"An hour or two will make no difference," Violet said severely. She picked up Sarah's bag. "Follow me, Miss."

Sarah followed, up stairs that climbed round a central stairwell. From above, the black and white diamonds of the hall floor shivered into each other. Sarah looked quickly away.

Her bedroom was under the roof, looking out onto the sky high above the trees of the garden.

Violet put Sarah's bag down on a chair. "Lunch will be in ten minutes," she said with a sniff, and disappeared.

Unpacking her bag took Sarah very few minutes. She tidied herself as best she could – the water in the pretty flowered jug on the wash-stand was cold – before going downstairs.

Lunch was frugal. Aunt Bessie pecked at her food and Sarah, anxious not to appear greedy, was reluctant to ask for more. Besides, she was kept busy answering Aunt Bessie's questions. Aunt Bessie wanted to know everything about Sarah, her sisters, her mother, Hillcrest – the questions never stopped.

"I always rest for an hour after lunch," Aunt Bessie said, rolling her napkin into a narrow cylinder and pushing it into a silver ring. "One needs little naps at my age, you know, dear. Violet thought it would be a good idea for you to lie down too. I expect you're tired after your journey."

In her bedroom Sarah looked down on the chintzy bedspread and wondered what to do. Was she expected to undress and get into bed properly? Take off her skirt and lie under the bedspread? At last she removed her shoes and lay on top of the bed, hands under her head, watching the reflection of the sunlight flickering over the ceiling and listening to the strange sounds of city life. She even thought, as she drifted into a light, disturbed sleep, that she could hear the roar of wild animals.

Violet woke her with a cup of tea. "When you're ready, Miss Elizabeth will take you into Miss Ellison," she said. She frowned. "I hope you haven't creased the bedcover."

Aunt Maud's room was dark and stuffy, the sash window tightly shut. A roll of material had been pushed between the frame to keep out the draughts. The curtains were pulled halfway across, leaving a thin rectangle of light. In the grate the coals were piled high, quivering with orange and blue flame. A large mirror hung above the mantelpiece like a painting, reflecting the room ('When did you last see your father?' Sarah thought – 'When did you last visit your great-aunt?'). It repeated the reflections in a smaller mirror on the opposite wall so that the room appeared to go on for ever;

hundreds of little tables laden with pill boxes and bottles of medicine, dozens of images of Sarah standing nervously by Aunt Maud's bed.

Aund Maud stared up at Sarah from the mound of pillows with bright, beady eyes. The yellow skin was stretched over the bones of her skull. Her mouth caved in, barely visible between the long, craggy nose and pointed chin. She looked like a wolf – Red Riding Hood's grandmother.

"So. This is Laura's girl." She pronounced it 'gel'. "Don't stand there with your back to the light. Let's have a look at you."

Her hands gripped the sheet, fingers curling over like claws, pigeon's claws.

"H'm. There is a touch of your mother about you, I suppose. Poor, silly gel."

"It's Frances who's like Mother," Sarah said, nervous, uncertain what to say, not used to being with invalids.

"Frances?"

"Oh, Maudie, you know Frances," Aunt Bessie said quickly. "She's the eldest – the one who writes to us at Christmas."

"Yes, yes. Don't fuss me. I understand Bessie's going to take you out – show you the place." She waved a claw feebly in the air. "Might as well let her have a look at the school, Bessie. Though I don't know what things are coming to these days. Gels going to school indeed. Governesses were good enough for me."

"You mustn't mind your aunt," Aunt Bessie said as they set off down the street. "She's a little touchy sometimes after her afternoon nap. She'll be better this evening."

"I don't mind," Sarah said. "Some people still don't think girls should go to school. Sir James doesn't. His daughter had a governess like Aunt Maud – did you have a governess, too, Aunt Bessie? She's coming out next year and then she'll get married."

She jumped over the cracks between paving stones. Only

railings separated the houses from the pavement; there were no front gardens at all. Behind the railings steps dropped steeply down. "Mr. Mackenzie says I'll go to university first."

In the basement, kitchen maids laughed and talked, polishing silver, preparing tea.

"I'm sure that'll be very nice, dear," Aunt Bessie said.

A group of cabbies waited for customers at the edge of the Downs. The warm, ripe smell – of horses, hay, full nosebags – reminded Sarah of home. So did the raucous cawing of rooks in the trees. In the distance she heard that other sound.

"What's that noise, Aunt Bessie? A sort of roaring."

"I expect it's the lions, dear," Aunt Bessie said comfortably. "The Zoological Gardens are not far away. We often hear the lions. Military bands too, sometimes. I do like military bands." She sighed. "We used to go to the Gardens often before Maudie was ill – to see the flowers of course, not the animals."

"I'd rather see the animals," Sarah said. "I've never been to a zoological gardens."

"We'll go together," Aunt Bessie promised, patting Sarah's arm. "Won't it be fun? We'll be able to do all sorts of things when you come to stay, all the things I haven't been able to do since Maudie's been ill. Oh, I am looking forward to it! Here we are, dear, here's the river."

Far below them a pale ribbon of water lay between broad banks of brown mud. Across the bridge, suspended over the void like a pen and ink drawing, the trees on the far bank were turning colour, browns and yellows against the evergreens. In the crisp autumn air nothing moved. Sarah looked past the stunted bushes clinging precariously to the rock face and tried to picture paddle boats at high tide, slowly steaming out to sea, towards the Quantocks' coastline and Minehead.

"I remember the day the bridge was opened," Aunt Bessie said. "It was a public holiday. There were fireworks and a band. I was just a girl then. Such fun it was."

They walked home through the school, their footsteps

echoing on the empty drive. Girls and teachers had left for the day. The buildings had a forlorn air, as if holding their breath, needing people to bring them to life. Sarah's hands grew clammy at the thought of tomorrow.

That evening Aunt Bessie and Sarah sat with Aunt Maud. Violet had pulled the curtains and stoked the fire. The picture in the mirror was dark, like a Victorian painting, as they sat in the gaslight and Aunt Bessie talked about relatives Sarah had not known existed.

"There were four of us," Aunt Bessie said. "Maudie was the oldest, of course. Then came Herbert — your grandfather — and Willie . . ."

"The black sheep," Aunt Maud said.

". . . who went to Canada." Aunt Bessie's cheeks were flushed. "I came last of all. I was the baby of the family." She smiled at Sarah. "Like you."

"Show her the photograph album, Bessie," Aunt Maud ordered and when Aunt Bessie left the room said triumphantly, "That's got rid of her." She peered at Sarah with beady eyes. "Want to go to school, do you, gel?"

"Yes, please," Sarah said.

"That's all right then." Aunt Maud nodded. "Glad to have you. Be like having Laura about again, I daresay. Though I don't know that that's a good idea. No running off to art school, mind."

"No art school would take me," Sarah said. "I can't draw at all."

Aunt Bessie returned with a large leather album. Sarah turned the pages, seeing stern, stiff men and tightly corseted unsmiling women. She thought how extraordinary it was to have been related to these people for so many years without knowing anything about them, and felt a delightful warm sense of belonging. Everyone she knew had relatives — the Mackenzies, Jessica — and now she did too.

"Are there any photographs of Mother?"

"Herbert told us to burn them after she had run away."

"We didn't, of course," Aunt Maud said. "They're in the top drawer of the desk over there. Fetch them, child. In the blue box." And as Sarah came back to the bed with the box Aunt Maud said, "Such a foolish boy, Bertie."

"But it's Frances!" Sarah exclaimed, picking up the first photograph: Frances in old-fashioned clothes, ready for a fancy dress ball at the Manor perhaps.

"No dear, those are all of your mother."

She looked at them slowly, one by one, wondering. Years ago Mother had sat in a chair and stared at the camera; the photographer had pulled out the shutter and there Mother was, caught for ever in black and white on a piece of stiff card.

Sarah had never thought of Mother as a child. Mother had been – Mother. What had she been thinking, Sarah wondered, as she stared at the photographer? The thoughts had gone, Mother had gone. Only the image remained.

"Are there any of Father?"

"No, dear. That's the last photograph, the one you're holding. It was taken just before she left."

"Ran away," corrected Aunt Maud.

"Why did she run away?"

"She quarrelled with her father – your grandfather. He wouldn't let her do what she wanted. They were both ... obstinate."

"Pig-headed, both of them," Aunt Maud croaked from the bed.

"Oh, Maudie, you mustn't say that. They couldn't agree about certain things, that was all. So she ran away."

"Wanted to go to art school. Can't blame Bertie for not agreeing with a nonsensical idea like that. Stupid thing was, she ran away, couldn't go to art school anyway. No money."

"So then you see she had to work. That was the trouble, the man she went to. Herbert wouldn't take her back after that." Aunt Bessie blushed. "Don't misunderstand me, dear. It was quite proper. Nothing like *that*! It was ... just ..."

"Spit it out, Bessie," Aunt Maud said. "No need to be mealy-mouthed."

"Well ..." Aunt Bessie leant forward and said in a hushed voice, "he was ... a socialist. They all were, the people she worked with."

"Frances says Gabriel's a socialist," Sarah said.

"Gabriel?"

"Gabriel Mackenzie. Mr. Mackenzie's son."

"Oh, I don't think so, dear. Such a nice man, Mr. Mackenzie. I'm sure his son couldn't be a socialist. You must have misunderstood Frances."

Sarah stared at her aunt. "Do you know Mr. Mackenzie?"

"Of course we know him. He came to see us when your mother died. And again last month to discuss your going to school."

Sarah looked down at the photograph she was still holding and thought of Mr. Mackenzie in this house, sitting in this chair perhaps, talking to them. She wondered what he had said. Had he asked them to take her? Or had they suggested it themselves? "Did Frances come too?"

"Told us he wanted to keep Frances away from the school," Aunt Maud said. "Sounds an interesting gel, Frances. She'll come up with you when you start; settle you in, he said. We'll meet her then."

"Tell me more about Mother. When did she come back?"

"She didn't."

"Never?"

"Herbert – your grandfather – wouldn't have her back until she'd begged his forgiveness. She wanted him to admit that he'd been wrong. They never even wrote to each other."

"You're wrong there, Bessie," Aunt Maud said. "She wrote and told him when she got married."

"So she did. She didn't say she was sorry, though, and when he wrote back he told her he was glad she realised he'd been right. That was the end. She never wrote again. Nor did he."

"Bought her the house, though, didn't he?"

"Home, do you mean? Hillcrest?"

"It was a wedding present, dear," Aunt Bessie said, "the nearest Bertie got to saying he was sorry. He thought she could live there when your father was at sea and let it when he was at home. That's why he bought a house where he did, halfway between Plymouth and Bristol. Convenient for them both, you see."

"He was sharp, was Bertie," said Aunt Maud. "Halfway, he said. Much nearer Bristol if you look at a map."

"Did he ever know about us?"

"Oh, yes. She wrote to Maudie and me every Christmas. Never to Herbert, but she knew we'd tell him."

"And they never saw each other again?"

Aunt Bessie shook her head. "We sent a telegram when he was taken ill. She came at once then, but it was too late."

"At least she was in time for the funeral," Aunt Maud said.

"What comfort was that to poor Bertie?" There were tears in Aunt Bessie's eyes. "It was so stupid. They wanted to make it up, both of them; you know they did. She should have ... but she was so obstinate. She'd always been obstinate, even as a child. She'd never admit that she'd made a mistake."

There was a knock. Violet stood in the doorway, the shadows from the lighted candle she held making a skull of her head. "Time Miss Sarah went to bed."

"Oh, Violet, do you think so?" Aunt Bessie sounded disappointed. "Yes, I suppose you're right. Off you go then, Sarah dear." She gave her a hug and added shyly, "It's so exciting having you here with us, after all this time. I hope – I know we're dull old sticks – I do hope you won't be bored."

"Don't be silly, Bessie," Aunt Maud said. "She'll have school to keep her occupied."

She lay in bed and listened to a distant clock chime the hours and half-hours. Her eyes pricked with tiredness. Her head throbbed from the unaccustomed heat and stuffiness of Aunt Maud's room. Her whole body ached. The maids creaked up the stairs to bed and giggled quietly in the room next door for what seemed an eternity. She tossed and turned, her mind full of strange images, buildings, streets, houses without gardens, the bridge hanging between the cliffs.

She told herself that it was lack of her bedtime cocoa that was keeping her awake, and tried not to think of Annie in the kitchen at home, so kind, so comfortable, so very different from Violet. Imagine Violet taking the trouble to make a pile of sandwiches, 'just in case', as Annie had done this morning.

Sarah sat up. Of course – it was hunger that was keeping her awake, hunger causing the pain in her stomach.

She crept out of bed and across the floor. She eased the bedroom door shut and lit her candle, wincing at the sudden *phtt* as the match flared into flame. Shivering in the cold, she undid the package that she had forgotten in the excitement of the journey. Bacon sandwiches, cheese and chutney, the last of the tomatoes – oh bless Annie, dear, dear Annie, who dreaded hunger and always feared the unexpected disaster that might leave a traveller stranded.

Hunger satisfied at last, she was about to climb back into bed when she remembered Violet. If Violet frowned at creased bedcovers what would she say to crumbs in the bedroom? Collecting them together in the flickering candlelight was not easy and what was she to do with them when she had finished? In the end she wrapped them in a spare handkerchief and stuffed them in the bottom of her bag for the sparrows at home. The pain returned at the thought of Hillcrest.

She fell asleep at last, and dreamed strange fitful dreams: of a deserted, echoing building that should have been a school but was not; of Mother and Frances who in some mysterious way had become one person, arriving late, too late for school, too late for forgiveness, too late for anything but a funeral.

Chapter Thirteen

It was as well that excitement had taken away Sarah's appetite next morning. The crustless toast at breakfast was thin and brittle, and there was not much to the marmalade but pale jelly. She thought wistfully of chunky pieces of fruit suspended in the solid jelly of Hillcrest marmalade and wondered whether to offer Annie's recipe to the aunts. Nervousness of Violet, standing like a warder at the sideboard, kept her silent.

She walked sedately to school by Aunt Bessie's side, her calm exterior hiding the turmoil within.

"I believe the school is — well, rather academic," Aunt Bessie said, in disapproving tones. "I would have thought . . . but I expect Mr. Mackenzie knows best."

There was no sign of life, not a person in sight. As they approached the ornate, turreted building they heard the distant singing of a hymn, sounding strangely thin without the deeper tones of male voices.

They were early. She sat with Aunt Bessie in the large, airy entrance hall waiting for prayers to end. Medici prints hung on the walls. Wide, shallow steps led up to the first floor. Through the round stairwell, through a skylight far above, she could see the sky. It was her favourite sort of day, bright, crisp, autumnal. At home, Gwen would be out in the garden, digging up the dahlias, clearing leaves; Frances already at her easel in the stables; Julia and Annie starting on the housework. How odd to think of the world of Hillcrest continuing as usual while she sat here waiting for her new life to begin.

The entrance hall filled suddenly with a vast quantity of girls, moving swiftly through in orderly crocodiles, their footsteps clattering on the tiles. As suddenly the hall was empty. Desk lids banged. Chattering voices were quickly silenced.

A young woman came down the stairs to conduct Aunt Bessie and Sarah to the headmistress's room on the first floor.

Sarah sat primly on the edge of her chair, remembering Mrs. Mackenzie's precepts — knees together, feet tucked underneath, gloved hands clasped on lap — while the headmistress and Aunt Bessie discussed hours and timetables and homework. There was a hockey pitch marked out on the grass outside. She had forgotten games were played at school. Perhaps she would be able to learn tennis at last.

" . . . not easy coming in the middle of term." The headmistress was talking to her, smiling with kindly eyes. "Particularly if you have never been to school before. Your guardian and I decided that it would be best for you to spend a day with us, in what would be your class. We shall be able to see whether we like the look of you and you the look of us. Have you any questions? No? Then the secretary will take you to your class."

Sarah felt momentary panic when Aunt Bessie departed with an encouraging smile and the promise to fetch her home for lunch. She followed the young woman again, this time along seemingly endless corridors, down stairs and along more corridors to the cloakroom. Would she ever be able to find her way around?

The cloakroom was dark, smelling of worn footwear and damp wool. There were pegs for coats and lockers for shoes.

"No talking in the cloakroom or the corridors," the secretary said as she led her down yet another corridor to what was to be her classroom.

Twenty faces turned as she entered; forty eyes stared. At least the mistress seemed to expect Sarah. She showed her to an empty desk and rebuked the class for not getting on with its work. "It is ill-bred to stare. You will have an opportunity to get to know Sarah at break. Kindly return to your exercise."

She was as thin as a pencil and as rigid. Her face was sallow, severe, the little mouth pursed up as if about to reprimand but she made several jokes during the lesson and there was a twinkle in her brown eyes. Sarah, nervous at first, began to enjoy herself. The work – it was Latin – was easy. She knew all the answers, although she was careful not to say so, speaking only when spoken to.

At the end of the lesson, announced by a jangling bell that reverberated round the building long after it had stopped ringing, the mistress asked Sarah to come outside.

"Now," she said, "that was much too easy for you, wasn't it?"

There were prints of Dutch interiors hanging on the walls of the corridor. One was the same as the picture on the landing at Hillcrest: a woman playing a spinet.

"I did know it all," Sarah said.

"Yes, I thought so." She sighed. "I was told that you had had more of a boy's education than a girl's. How long have you been learning Latin?"

She tried to think back. "I can't quite remember."

"What about Greek?"

"About two years, I think. Properly, that is."

"I see." Her eyes twinkled. "How long – improperly?"

"Well ... all the time, I suppose. The person I had lessons with was a lot older than me, you see. We were in the same room so I picked it up from him."

"As you presumably did with Latin?"

"Yes."

"The girls in your class start Greek next year. A few that is, not all. You'll be very far ahead; as you are in Latin of course. We shall have to think what to do."

"Do you teach Greek, too?" Sarah asked timidly.

"Yes, I'm in charge of classics. We shall obviously be seeing a great deal of each other in the future. Run along now, you mustn't miss your next lesson. And don't worry, we'll work something out."

The following lesson was history and after that English. The class was studying Keats's sonnets; today it was the turn of *On first looking into Chapman's Homer*. Sarah had not read it before and thought it appropriate: '. . . a new planet swims into his ken'.

Next time the bell rang the girl in front of her said, "It's break now. You'd better follow me," and led her down corridors to a room where milk and a bun were dispensed to every girl. The drink was an unexpected pleasure after the chalky atmosphere of the classroom.

Break was spent outside, surrounded by girls from her class who looked at her as if she were a stranger from a foreign country.

"What do you think of school?"

"It's very nice."

"What do you *really* think?"

"It's a bit . . . strange."

"Strange? Why do you think it's strange? In what way strange?"

"Well . . . don't you *talk* to the teachers – mistresses – at all? In class, I mean. Don't you discuss things with them?"

They stared at her as though she were a Hottentot. "Of course not. They're grown-up."

"Yes, but . . ." She talked – she liked talking – to grown-ups. Except for Antony and Jess – and now that he was in the army Antony probably counted as one, too – all her friends were adult.

"Whatever would you discuss anyway?"

"Well . . . your opinion. What you thought of something."

The girls giggled. "They'd be frightened to do that. Think

what we might say. 'Boring old Shakespeare!' Miss Allison wouldn't like that, would she?"

"You don't think Shakespeare's boring, do you?"

"Don't you?"

"I don't know. I haven't read any yet."

Mr. Mackenzie thought that she was still too young to appreciate Shakespeare's work. All the same she had frequently acted out scenes down in the spinney with Antony and found them exciting, spine-shivering sometimes. Yet they found Shakespeare boring.

How long did break last? She was penned in, unable to escape. Other girls walked up and down the drive. Up to the gate, touch it with one foot, turn and down again, talking. Didn't you get bored doing that day after day? Couldn't you escape to a quiet corner with a book?

"My brother's in France," someone boasted. "I bet you haven't got anyone in the war."

"I've got three," Sarah said quickly.

"Brothers?"

She hesitated. "Yes." They were as good as brothers; she could call them brothers surely.

They looked at her with respect.

"In France?"

She could not take the fibs that far. "Not quite; not yet. They'll be going soon though."

She was rescued by the bell. The girl who had helped her before said, "My name's Kate. We have French now and then arithmetic."

The French lesson was terrible. She had never learnt French; could understand nothing of what was said. What was worse, she had difficulty understanding Mademoiselle Gautier's English. The class giggled. Her cheeks burned. She knew that they must think her stupid.

The arithmetic lesson was little better. The other girls were quicker at working things out; they finished each exercise before she was halfway through.

There were other difficulties. At the Rectory she worked at a table. She found writing on the slope of the desk awkward; the desk was cramped and her shoulder blades in the wrong place for the chair. It was a relief when the bell rang at the end of the morning.

Kate walked down the drive with her. "Do you have far to go – where do you live? Gracious, is that your mother? She looks awfully old."

Kate's mother was pretty. And young, years younger than Mrs. Mackenzie.

"Well, dear?" Aunt Bessie watched Sarah anxiously.

"I don't know," Sarah said uncertainly. "It's very noisy." Her head ached from the sound of bells, clattering footsteps on bare boards, banging desk lids and high-pitched voices.

She sat in the over-furnished dining-room, listening to the gold clock on the mantelpiece ting away the dinner hour and let Aunt Bessie chatter through the meal. The thin slices of meat and mushy, over-cooked vegetables tasted of chalk and were unaccountably difficult to swallow.

"You don't need to come back with me, Aunt Bessie," she said after lunch. "I know the way."

Aunt Bessie hesitated. "Are you sure, dear? I do like to take a nap, but only if you don't mind ..."

Sarah's footsteps slowed down as she reached the school. It required a real effort to go through the wrought iron gates. Someone called as she walked up the drive. Kate sat among a group of girls on the low wall under the copper beech. Sarah recognised faces from her class. Someone asked: "Haven't you ever been to school before?"

She hated the thought that they had been discussing her. "No."

"You must have had lessons."

"Yes."

"Who taught you?"

"My mother at first. Then the Rector where I live. He taught my sisters, too."

"Are they coming here?"

"They're too old."

"How many sisters have you got?"

"Three."

"So there are seven of you altogether?"

"Seven?"

Their eyes accused. "You said you had three brothers."

"Oh. Yes, of course. Seven." Or should it be eight? What about Lucy?

"Are you telling the truth?"

"Of course."

"Are you pi?"

"I don't know."

They giggled. "You don't know what it means, do you?"

She shook her head.

"If you come from a rectory you must be pi," Kate said, tossing her head. Her hair barely covered her ears. Sarah had never seen a girl with short hair before and was fascinated by it. How cool it must be, how easy in the morning. No endless brushing, no difficulty getting the plaits tight enough at the top, no ribbons slipping down. When Kate tossed her head, her hair flew up before coming down to settle neatly on her ears and into the nape of her neck. How wonderful to be Kate, to have short hair, to understand what strange words meant, to know one's way about. Sarah looked at the girls in front of her, self-possessed, immaculate in their pleated gym-slips and white blouses. What sort of clothes would Frances expect her to wear? Frances's old school uniform, the wrong colour, handed down through Julia and Gwen?

The afternoon was worse than anything that had gone before: the entire afternoon given over to art.

Sarah, expecting someone like Frances, was astonished by the art mistress. Nothing to look at at all, poor soul, Annie would have said. Plain, in other words. But she had a certain presence despite her ordinary appearance and in no time at all had quietened the jostling, chattering girls and got them sitting at the tables, silent and attentive.

"Last week, you may remember, I asked you to paint the school building. I thought, foolishly as it turned out, that you would have no trouble drawing it from memory. You are in and out every day, after all. How wrong I was! What extraordinary buildings I got! Today you will correct and improve. I'll come to you in a moment, Sarah, when I've returned these drawings. Sarah's sisters, girls, are professional painters so we shall be interested to see what Sarah can do."

Why did she have to say that? Sarah wondered, scarlet-faced, as the class turned and stared. What would they say when they saw her work?

"Are your sisters really artists?" her neighbour asked; she could hardly have sounded more surprised had she been told they possessed two heads apiece. "What do you think of my picture?"

Her drawing was awful, worse than anything Sarah might have done. Sarah didn't know what to say. "I think it's the perspective that's wrong," she said at last.

The artist glared. "What do you know about perspective?" she asked and removed herself and her drawing to a spare desk at the far side of the room.

Sarah was crushed. Frances never took offence like that. And it was true, the perspective was wrong, dreadfully wrong. Didn't the stupid girl know that lines always met on the horizon?

The mistress made her way between the tables to Sarah. "Asking you to paint the school building would be a little unkind, wouldn't it? What would you like to do? You don't want to tackle something you can't finish this afternoon. How about a portrait of one of your sisters?"

"I'm not very good at portraits," Sarah said nervously, not sure how you painted someone who was not there before your eyes.

"Your home? Something you know well?"

"Yes, I'll do that."

She busied herself with the pencils. Life was unkind: why did it have to be art today of all days? She wondered what she should paint, summoning up pictures of Hillcrest in her mind. The greenhouse at the end of the verandah, shadowed and semi-tropical, full of Gwen's orchids and black grapes hanging from the vine? The corner of the stable block where the clematis twisted in such profusion through the shrub rose that both had fallen away from the wall and lay in a waterfall of colour? The images were there behind her eyes but she knew that she would never be able to put them down on paper.

She drew the stable block at last. She could picture it with such intensity that she almost felt the grass under her feet, heard the hens clucking as they pecked at the earth floor under the staircase up to Frances's studio and the rooks cawing in the trees of the copse beyond the wall. She could show none of it on paper, nor could watercolour, the only paint available, adequately portray the thickness of the thatch, the solidity of the sandstone, the way the yellow irises pushed up through the encroaching grass.

The art mistress went round the class, making comments, giving advice. "That's nice," she said, looking at Sarah's picture. "Very modern."

Sarah took her words as criticism and flushed.

"You may take it home if you like."

She shook her head. Who would want a painting like that, pale, insipid, a mere imitation of reality? Imagine Frances's scorn when she saw it.

In the cloakroom at the end of the afternoon, Kate said, "When do you start?"

Sarah fumbled with her shoes. "I don't know."

"I'll keep that desk free for you if you like. Come on – aren't you ready yet? You do take a time." She ran down the drive in front of Sarah, shouting back, "Shall I? Keep the desk, I mean?" Her hat swung from her fingers, her hair bounced against her neck. The young woman was waiting again at the gates. She put her arm round Kate and they walked away together, laughing.

The shadows were lengthening as Sarah walked slowly back to the aunts. The days were drawing in: it would soon be winter. The shadows of the houses lay in black blocks across the pavements and over the tarmac. The air was clear, crisp with a hint of frost. As Sarah reached the square two nursemaids came out of the gardens. Manoeuvring their prams over the step, engrossed in conversation, they did not notice Sarah slip past them through the gate.

The gardens were immaculate. No disarray, no small patch of wilderness. Not a weed to be seen, no sign of even a clover leaf among the close-packed blades of grass. Shrubs had been put in discreet groups, trees lopped when they had grown too big. Nature controlled by man. Empty flower beds lay like giant wedges of cake cut out of the lawns, the soil worked into a fine black tilth, waiting for the spring bedding. In the distance two gardeners were clearing the debris of autumn, sweeping the paths with slow deliberate movements. A spiral of smoke twisted lazily into the pale sky.

Sarah sat down. The iron struts of the seat made bars against her back. Life here would be ordered, proper, tidy. And public. The windows of the houses frowned over the square. There was nowhere to hide, no retreat at all.

She thought of Aunt Bessie, waiting to show her Bristol. She thought of her own dreams of school, the long years of make-believe dissolved in a few hours of reality.

She got up at last and walked slowly to the gate.

It was locked.

For a moment she would not believe it. Then she panicked. It was years since she had been frightened by wolves, but distant sounds reminded her of real animals near at hand. Who knew how weak were the bars that caged them in? Frantic, she rattled the gate, struggled to prise open the padlock, finally flung herself against the cold railings and wept.

Then she remembered the gardeners.

They looked at her blankly. "Shouldn't be here without a key."

She said humbly, conscious of tearstains, "I didn't know."

They went on sweeping. The leaves twisted lazily over in front of the spiky twigs of the besoms, red and yellows turning to maroons and browns. "Where do'ee belong, then?"

"I'm staying at number eight."

"Aye." Their voices were expressionless. Scutch, scutch went the besoms. The dead leaves crackled.

"Please let me out," she said, ashamed of the quaver in her voice.

There was little comfort at number eight. Aunt Bessie was crying. The kitchenmaid who opened the front door to her was in tears. From the back of the hall Violet glared with eyes small and cold with dislike.

"And where have you been, madam? Stop your snivelling, Mabel, and get back to the potatoes. Three times to the school and back I've been this afternoon, let me tell you. I've enough work of my own to do without going out looking for you."

"Don't be cross with her, Violet," Aunt Bessie begged. "Oh Sarah, where have you been? We've been so worried . . . what would I have said to Mr. Mackenzie . . . ?"

"I was locked in the square garden."

Aunt Bessie gave a moan. "Sarah, *dear*, not in the gardens on your own! Please don't tell your Aunt Maud." She blew her nose, patted Sarah's shoulder. "Still, you're home now, safe and sound. But you know, Sarah, I don't believe you're supposed to walk to school alone. Not at your age. It was very wrong of me. I should have walked with you this afternoon; and met you too."

"Can't I go anywhere by myself?"

"Dear, Bristol isn't Huish Priory, you know. The square isn't like Hillcrest. Things are different here."

"Yes," Sarah said. She felt sick with tiredness. "I see that."

In the attic bedroom she knelt at the window and pressed her nose against the glass. The sun had disappeared behind the houses across the square, leaving the gardens below in cool shades of grey, secretive, remote, guarded by black sentinels. There were bars, like the railings, outside her window, turned over. Vertical bars . . . horizontal bars. Bars to keep people out . . . bars to keep people in. Prison bars . . .

Chapter Fourteen

Sir James was pacing the platform when the train from Bristol drew into Taunton station. Sarah, worried about changing trains and the possibility of arriving at Barnstaple instead of Dunkery St. Michael, was so pleased to see the familiar face that she could have hugged him.

"There you are, young lady. Had a good journey? Let's have your bag. When Mr. Mackenzie knew I was coming into town he suggested that I collect you. Thought it would save your sisters getting wet. What weather, eh?"

He helped her into the car and wrapped a rug round her legs.

"Hawkins volunteered, you know," he said as he got into the driving seat. "Damned inconsiderate. I told him so. Hawkins, I said, you're a fool. You're hopeless with a horse: what good do you think you'd be in the army? Wouldn't listen." The car moved, jumped as he double declutched, and leapt forward. "There's no call for the working class to go out and enlist, you know. Besides, this machine's beyond me. Give me a horse any day."

The car swerved round a corner and mounted the pavement. A flushed face appeared at the window and swore. Sarah clutched the bag on her lap and tried not to think about accidents.

"Look where you're going, you stupid man!" Sir James shouted at the window.

Usually so kind, so even-tempered – was Sir James always like this in a car? She was relieved when they had left the town and were driving along the empty Minehead road. She sat in a state of dreamy contentment, wrapped round by the pleasant aroma of leather, old tobacco and wet tweed, mesmerised by the rain that beat against the windscreen. Through the steamed-up glass the Somerset landscape appeared shadowy and remote: another country, another world. The leafless trees loomed silently up like ghosts and as silently faded away into the mist.

There was no sound or sign of life when they reached the village, but the steady trickle of water dropping from the cottage roofs and running away down the side of the street.

The car came to an abrupt halt on the Manor drive, scattering gravel onto the grass.

"Do you mind walking from here?" Sir James said. "The rain's let up a bit now. Don't fancy turning in the mud of your lane, to tell you the truth. Hawkins might manage it, but I'd rather not try."

"Of course," Sarah said. "Thank you for meeting me."

From the gates of the Manor only part of Hillcrest was visible over the brow of the hill. Oblivious of the rain, Sarah stopped and gazed at the slate roofs, the glimpse of apricot walls, overwhelmed with happiness.

The chestnut leaves which two days ago had been lying in layers of yellow and browns on the road had had their colour washed out of them by the rain and now lay black and lifeless under her feet. During her absence the last leaves had come away from the virginia creeper by her bedroom window and drifted onto the flowerbed beneath.

She saw it all as she ran up the hill in the rain and burst through the front door. The hall was darker than she remembered and smaller . . .

"Oh, love," Annie exclaimed, "you're soaking. You'll

catch your death of cold. Off with that coat – quickly now. Go and sit by the fire. Have you had something to eat? Did the aunts give you anything for the journey?"

"The aunts don't eat," Sarah said and giggled.

She looked round the familiar kitchen, warm and glowing in the lamplight, while the rain beat down outside on the cobbled yard and lashed against the window. She remembered how she had felt shut out from the basement kitchens in Bristol, and sighed with relief. She was home, back in her own family – Annie lifting the earthenware pot out of the oven, cheeks flushed from the heat of the range; Gwen with earth under her fingernails and plant labels sticking out of the pocket of her hessian apron; Frances in her painting smock, with ultramarine on her nose; practical Julia, fetching plates and a mug from the dresser; and Elsie coming in from the scullery to gaze in awe because Sarah had travelled all the way to Bristol and back on her own.

She wanted to hug them all, to sing and dance round the table. Instead she stood speechless, a foolish grin stretching from ear to ear.

"Tell us about it," Gwen said, "right from the beginning. What was school like? When do you start? Oh dear, it'll be horrid without you."

"I'm not going," Sarah said, her smile gone. In the train she had practised breaking the news, firmly but gradually. When the moment came she could only do it baldly, abruptly. She said again, "I'm not going."

"Not going! But Sarah, why ever not?"

"I don't think it's a good idea, that's all. Don't let's talk about it."

They stared at her, silent and bewildered.

"Tell us about the aunts then," Julia said at last. "Are they very old? What are they like?"

So she told them about the aunts and the tall, thin house with its garden behind bars across the road, while Annie ladled stew onto a plate, thick pieces of meat, chunks of vegetables, carrots, onions, potatoes and rich golden gravy.

"Oh, Annie," Sarah said, breathing deeply. "It smells delicious. I'm starving. The aunts hardly eat, you know. You can see the pattern on the plate through the slices of meat they give you. Truly. I was so hungry last night I hardly slept a wink."

"No wonder there are shadows under your eyes then," Annie said tartly. "Proper peaky you look. Eat up before you fade away."

Questions were bound to come, Sarah knew that. For the moment she was home. It was all that mattered. She soaked up the gravy with crumbling hunks of fresh bread and no-one, not even Frances, sitting on the dresser watching her with brows drawn in a black line above her thoughtful eyes, not even Frances reproved her.

Later, when the rain had stopped, she went outside. The garden was very still, holding its breath as if a mere tremor would dislodge the drops of water still shivering on leaves and twigs. Rain had changed the colours of grass and paths and stained the pink sandstone wall with darker shades of red. She roamed round the garden, seeing things that she had always seen and never noticed, shrubs and trees growing unrestrained in jungle patches, fruit trees and clematis climbing together over walls, the wrought iron weather-vane silhouetted against a dark grey sky, delicate seedheads and pale skeletons of leaves left over from the exuberant growth of summer.

Under the trees rain still dripped steadily from the branches onto the fallen leaves beneath. She looked down on the glistening soil, the rotting medlar fruit with their curious cruciform scar, and breathed in the sweet smell of decaying vegetation, the brown woody atmosphere of autumn. Her time in Bristol was like a scene in a book that she had read and quickly discarded. Hillcrest was here for ever.

She could only delay the moment when Frances would demand explanations. Tomorrow, she hoped. Not today. But that evening Frances came to her room when Sarah was getting ready for bed.

Sarah sat at the dressing-table, avoiding Frances's eyes in the mirror, and brushed her hair with a vigour that would have astonished Annie had she seen it. She talked, quickly and nervously, about the suspension bridge, lions in Zoological Gardens, the street lights – anything to delay the questions. She brushed and brushed, until her hair crackled and jumped. Frances never moved. Frances had the gift of absolute stillness, like a cat watching a bird; it was one reason why she had so little patience with people who fidgeted when posing.

Sarah gave up at last and climbed into bed. "Gracious," she said, stretching her jaw in the biggest yawn that she could manage, "I am tired. Good-night, Frances."

Frances stood up and shut the door. "Now perhaps you would like to tell me . . ."

Sarah looked at her wide-eyed. "I don't know what you mean."

"Of course you do. What's this about not going to school? I thought it was what you had always wanted."

Sarah fingered the seams of the patchwork on her quilt. It was true. She had always dreamt of attending a proper school. She wondered now how she could have been so foolish. "I didn't know what it was like."

"You can't judge from one day. I'm sure you'll like it when you get used to it."

"Oh, Frances," Sarah said in great fright, "don't send me, please don't send me. I couldn't bear it."

"Why not? I wouldn't force you, you know that, but I want to do what's best for you. What went wrong? What happened?"

"Nothing. It was just . . ." If she was not sure herself how could she explain to Frances? "We hadn't done the same things. They hadn't learnt Greek and I don't know French."

"We knew that before you went. Mr. Mackenzie discussed it with the headmistress; she thought there was nothing that couldn't be worked out somehow. Mr. Mackenzie thinks that you need to work with other people. Now Antony's gone you're on your own. He's not sure that that's a good thing."

Cold fingers clutched at Sarah's stomach. "Doesn't Mr. Mackenzie want to teach me any more?"

"I don't know. We haven't discussed it. We thought you'd want to go to Bristol."

"I'll work very hard if he'll let me stay. Please let me stay. I won't be any trouble, I promise."

Frances came over and sat on the bottom of the bed. "How often are you any trouble?" She smiled, rather sadly. "I do worry about you, mouse."

Sarah stared at her in astonishment. "Worry? About me. Why?"

"You're so much younger than the rest of us. We've got each other to discuss things with, argue things out. I wish I knew what Mother would have done. I don't want you to be lonely, but you've got nobody of your own age."

"There's Jess."

"Jess starts at Nether Stowey in the spring. Mrs. Mackenzie's found her a place as a nurserymaid. You won't have Jess after April."

"I'm not lonely." She was not, she was sure she was not. And yet ... was that what she had felt yesterday, among those girls? Surrounded, yet apart, separated by an invisible, inexplicable barrier. Was that what loneliness was?

She frowned at the candles on the table and searched for words to explain, to make Frances understand. For now she knew why she didn't want to go to school. She was different. She didn't look like the girls she had met, act like them, think like them. However much she might want to be like them — and she did — she knew she never would.

She was silent. How could Frances understand? For Frances was different, too. Frances said things that shocked people, outraged them. People talked about her behaviour. But Frances didn't care. All that mattered to Frances, all that had ever mattered, was her work. Sarah tried to put Frances into yesterday's setting and failed.

"Did you like school?" she asked.

Frances shook her head. "I hated it, every minute of it. Don't worry, if you really don't want to go, I won't make you." She smiled at Frances, teeth catching her lower lip. "I'll tell you something that no-one else knows, so don't you go blabbing it out. If Mother hadn't died when she did I'd have been asked to leave."

"Leave school?" Sarah's mouth dropped open. "Expelled, do you mean? How do you know?"

"The headmistress told me so when she broke the news about Mother. She said perhaps it was just as well, because I was very unsatisfactory and she'd been about to ask Mother to take me away. I can see her saying it now: nasty yellow face and mean little eyes. Heavens, she was a dreadful woman! Fancy saying that to somebody who'd just been orphaned. I was so angry I could have killed her."

"But Frances, why?"

"I don't know. I think there were all sorts of reasons. I was insubordinate, she said. Well, I've never agreed with rules just for the sake of rules. And I didn't fit in. The school was for daughters of naval officers, you know; it was there to churn out future wives for the navy. I couldn't care less about the navy or naval officers. All I wanted to do was paint, and naval wives don't paint. Probably it was as simple as that."

"Some wives paint. Mother did."

"She didn't, you know. Oh yes – before she met Father and when they were first married. Then we came along; she was trailing round the country after Father. Painting wasn't possible. I daresay she could have carried on with pleasant little water-colours, the sort of thing Lucy does, but that wasn't what she wanted."

"When Father died . . . "

"She tried, I know she did. But she hadn't painted seriously for – how many years? Fourteen or fifteen. You can't pick up something like painting where you left off years before. I'm sure that's why she died really. Her painting didn't matter when Father was alive. When he died and she tried to take it up again, she found she couldn't. Not like she used to do anyway. I found her in front of her easel sometimes, crying because she couldn't achieve what she wanted. You never can, of course, but I suppose it's worse if you've been good once. She'd lost Father, she couldn't paint – she'd got nothing left at all."

"She'd got us."

"She'd got you. Gwen and Julia perhaps. Not me."

"What do you mean?"

"We had terrible rows, Mother and I. Annie says we were too much alike. I did love her but we argued all the time – always about painting. She wouldn't let me paint. She wanted me to be like everyone else: get married, have children. She'd been happy doing that. It was trying to paint afterwards and not succeeding that made her so miserable. But you see I think she was wrong. If she'd never married, if she'd painted all her life, she would never have lost the touch. Perhaps she realised that in the end. She did leave me her paints when she died."

The candle on the table behind Frances outlined her profile against the darkness with a soft, yellow light.

"Professor Tonks always said that men and women were just the same when it came to talent. Until they married, that is. Then the women became weighed down with worries and work and things like that and their painting tailed off. I thought of Mother when he said that. That's when I decided never to get married."

Sarah stared. "Not ever?"

Frances smiled. "Not ever."

"But ..." What about Gabriel? Sarah wanted to say. She said nothing.

"We've wandered off the subject of school, haven't we?"

Frances said at last. "Can you give me any good reason for not wanting to go?"

"It wasn't only the school," Sarah said slowly. "It was the aunts, too. I did like them, but ..."

"They want to have you. Mr. Mackenzie's sure of that. They aren't just doing us a favour, or taking you out of a sense of duty."

"I know," Sarah said. She thought of Aunt Bessie looking forward to her arrival, planning visits to the Zoo again, waiting to do things with her that had been impossible during the years of Aunt Maud's illness. She felt a pang of regret, of remorse, but not even Aunt Bessie could make her live in Bristol. "I don't know how to explain. It was so ... so tidy. You couldn't take your shoes off." She thought of the few moments that afternoon, standing under the medlar tree with the water dripping off the trees. "I don't think I could bear to leave Hillcrest."

"Oh, well," Frances said, "that's different. If that's how you feel of course you needn't leave Hillcrest. It's how I feel, too."

"You spent all that time in London," Sarah said, surprised.

Frances laughed. "I'd have spent three years in Timbuctoo if the Slade had been there. All the same, I used to get desperately homesick. It wasn't the Slade, which was marvellous, so much as London and being away from Hillcrest. That first term was awful. If it hadn't been for Gabriel coming down from Cambridge and taking me out at weekends and trying to cheer me up, I don't think I could have stuck it. He kept on telling me to imagine his mother's face if I gave up." She hugged her knees, smiling. "Well, you can, can't you? And then at the end of term I used to sit in the train and grin like a Cheshire cat all the way home. Like you this afternoon, now I come to think of it."

"Is Gabriel really a socialist?"

"He's a Fabian."

"Isn't that the same thing?"

"No. They both want to change the way things are. The Fabians think it can be done peacefully. The socialists want a revolution. I think that's the difference."

"I'm glad he's a Fabian then. I wouldn't want Sir James's head chopped off."

"I don't think that's very likely." She stood up and stretched slowly, like a cat. "About Bristol. Think about it a bit longer. We don't have to decide tonight. You may feel differently in the morning." She smiled at Sarah. "It'll be nice for us if you stay. Hillcrest wasn't the same without you. I missed you, mouse." She bent down and kissed her – Frances, who rarely kissed anyone. Then she pinched out the candles and went lightly from the room.

Sarah remained propped up against her pillow, the image of the candle flames flickering still in front of her eyes. Slowly she slid down between the cold, cold sheets, stretching her legs out towards the warmth left by Frances.

Oh, Gabriel, she thought. Poor Gabriel. Do you know? Has she told you?

But it was not Gabriel who occupied her thoughts as sleep overtook her. Her mind was filled with the astonishing, the wonderful discovery that Frances cared, that Frances worried about her, that Frances had missed her . . .

1915

Chapter Fifteen

In a village the size of Huish Priory weddings were infrequent events, each one being a source of conversation for months beforehand. The fact that on this occasion the engagement lasted weeks rather than months or years meant that the gossip had to be all the more concentrated. After an afternoon spent with Jessica, Sarah was bubbling over with news to impart to her sisters in the evening.

"There's to be a guard of honour – soldiers from his camp, Jess says. Do guards of honour have swords or rifles, do you think?"

"Hay rakes, I imagine," Frances said dryly. "Unless things have improved recently."

"Imagine coming out of church under an arch of swords," Sarah said dreamily, and then stopped. "Suppose one dropped and chopped off your head."

"Oh, Sarah," Frances said. "Do stop talking such rubbish. Lizzie Roberts is a foolish girl and I hope you're never tempted to follow her example."

"Why's she foolish? Jess says . . ."

"They hardly know each other. When you think how long Bill Roberts and Mary were walking out I'm surprised that old man Roberts didn't insist on a longer engagement."

"He seems a nice enough lad," Gwen protested, "though I agree that she knows precious little about him. Still, I suppose things are different in wartime."

"They couldn't have a longer engagement," Sarah said, "because he's going to France. I think it's very romantic. And sad. They're only having two days' honeymoon before he goes to the front, Jess says."

"I hope Lizzie doesn't live to regret it," Frances said. "It's all very well being carried away by romance or patriotism or whatever you like to call it. The rest of one's life is a very long time."

"Lizzie's still going to live at home, after the honeymoon. Jess says . . ."

"Jess, Jess, Jess," Frances said irritably. "Don't you ever say anything?"

Sarah stared. "Of course I do. It's just – don't you think it's romantic too?"

"There's more to marriage than romance. Don't let people like Jessica fill your head with silly ideas."

Sarah said nothing, but she was hurt. Just because Frances didn't believe in marriage . . .

"By the way," Frances said, "I'm going away for a few days."

"When?"

"Tomorrow, as a matter of fact."

There was a shocked silence.

Gwen said, "But Gabriel's coming home tomorrow."

Frances went on sewing, her cheeks flushed in the firelight. "Yes, well. It's just one of those things."

"How long for?"

"Two weeks probably."

"Where?"

"I'll leave my address with Annie. I'll be with friends from the Slade."

"But Frances, how can you go away now? What about Gabriel? It's his last leave. He's off to France next week."

They stared at Frances who continued sewing, apparently unconcerned, though Sarah noticed that the stitches were getting bigger and bigger and the seam itself was crooked. Frances would have to do it all over again tomorrow.

"Gabriel'll be awfully upset if you're not here," Sarah said.

Frances looked up. "Time for bed, Sarah."

"It's not. It's hours to bedtime."

"It won't hurt to have an early night for once."

"Frances, *please*."

"Hurry up."

Frances had put on her most obstinate expression. Sarah knew that there was no point in arguing. Useless to seek help from the others, either. They were only concerned with Frances and her plans, gathering breath to protest, to argue and do battle.

They never let me say anything, Sarah grumbled as she climbed the stairs. They always send me away. It's not fair . . .

No-one came to say good-night. She lay forgotten in bed, listening to the rise and fall of angry voices downstairs; Gwen and Julia, Annie brought in from the kitchen, arguing, pleading. She lay in the dark and thought about Frances going away, not caring that Gabriel was leaving England, perhaps for ever. But all the time there was one picture dancing in front of her eyes – Frances, caught in the sunlight at the bottom of the churchyard steps.

Breakfast was eaten in silence. Everything that could be said had been said the previous evening and Frances's Gladstone bag, packed and waiting in the hall, told Sarah all she needed to know.

Frances, heavy-lidded, pale, looking as though she had not slept, picked at her food and kept her eyes on her plate. Not until they gathered in the hall to see her off did she look her sisters in the face, asking wordlessly for sympathy, knowing approval was impossible. "It's the best thing to do, really it is."

They were silent. It was Annie who burst out, "I don't know how you can do it, Miss Frances. To Mr. Gabriel, of all people. I'd never have thought it of you. Hard, that's what you are, hard as nails."

Frances's cheeks flamed. "Don't you dare say that to me, Annie. It's not true. I want to stay . . ." She stopped, picked up her bag and said wearily, "Give him my love. Tell him I'll write."

Even then, none of them thought that she would do it. It was impossible to believe that Frances, stubborn, independent as she was, could be selfish enough to let Gabriel come home and find her gone. But with a barely perceptible tilt of her head she stepped out of the dark hall into the sunlit front garden and set off down the hill towards Dunkery St. Michael and the train.

"Well," said Julia as the sound of footsteps died away. "That's that." She shut the door with a bang. "I suppose we should be glad that one person at least will be pleased. So pleased I daresay she'll break the news before he's even had time to step inside the door."

He came over to Hillcrest early the next morning. Sarah heard the familiar voice as soon as she came down the steps into the yard after letting out the hens, and peering round the scullery door saw him standing in the passage outside the kitchen.

"I'm sure I don't know, Mr. Gabriel," Annie was saying. "I'm no mind reader at the best of times and Miss Frances's is quite beyond me. Don't ask me what she was thinking of."

"She knew it was my last leave – I wrote and told her I was going to France. How could she go away?"

There was no reply from Annie but the steady chop, chop, of knife on chopping board.

He said desperately, "She must have left an address. Didn't she?"

Chop, chop, chop. The strong eye-pricking smell of onions reached out to Sarah in the scullery.

"Annie, please."

"It's more'n my life's worth to let you have it," Annie said. She sniffed hard. "I'm sorry, Mr. Gabriel, really I am. Now, if you don't mind, I've a lot to do."

The moment the idea came into Sarah's head she jumped out of the scullery without giving herself time to think and ran up to Gabriel.

"Frances said you could have a painting to take with you. She said you could choose." She pulled him towards her, away from the kitchen doorway and the sight of Annie's astonished face. "Let's go now."

He gripped her shoulder. "Did she really . . . ?"

"Gabriel, don't. You're hurting."

"Which one?"

"In the stables. You'll have to get the key. I can't reach."

He lifted the key from its hook in the scullery while Sarah waited for Annie to call out, to say that it wasn't true, that it was Miss Sarah's imagination again. Annie, still chopping onions, said nothing.

Gabriel walked quickly, grim-faced, along the path that ran up the hill beside the stable block. Sarah had to run to keep up, tripping over the hens that clucked and pecked in the long grass. A broody Rhode Island Red sitting in the stable doorway stared blankly into the distance as Gabriel took the steps up the studio – half staircase, half ladder – two at a time. He paused at the top, hand on the key in the lock, and looked down at Sarah.

"May I go in?"

Frances's studio had been made from the space between the eaves and the coach-house and storerooms beneath. In the days when Hillcrest boasted horses and a carriage, the stable-boy had lived there. Frances had cleaned it, whitewashed the walls, had a skylight put in the roof and furnished it with a few pieces of pale, unstained wood – bed, chest-of-drawers, table, a couple of chairs and a stool. The room was light and airy, the only colours the faded blue silk cover, relic of Commander

Purcell's time in China, thrown over the bed and the Arthurian mural on the far wall painted during what Frances called her pre-Raphaelite period.

The paraphernalia of her work lay about the room: her paint-stained smock hanging on the back of the door, stacked canvases, brushes jumbled together in a jug, plaster casts standing in one corner. The still life on the table, a patterned plate, china fruit and a piece of Bristol glass borrowed from the Mackenzies, had already been blocked out on the canvas resting on the easel.

The familiar smell, of paint and linseed oil, chalk and turpentine, filled the room, so reminiscent of Frances that Sarah became afraid. At any moment Frances might return, ask what they thought they were doing in her studio without her permission. What would she say if she knew they were here, looking at her work, touching her possessions?

Gabriel sat on the couch, staring down at his hands, in an attitude of utter despair. People talked of his black moods, Antony had mentioned his depression more than once, but Sarah had never known him other than affectionate, cheerful, teasing. She did not know what to say to this stranger.

She had learnt long ago that unhappiness was not confined to children, as she had once assumed, but she had always taken for granted that adult unhappiness resulted only from events over which there was no control, illness say, or death. It was hard to believe that a simple action – going away when it would have been as easy to stay – could produce such unhappiness; harder still to accept that it was her own sister who was responsible.

She sat down beside Gabriel. His army haircut left a band of pale skin between his hair and the tan on his neck. For some reason that white line upset her almost more than his present unhappiness.

"I have French lessons now with Madame Defosse," she said to break the silence. "She's a Belgian refugee. Mr.

Tasker's given her two rooms at Clay Court. Monsieur Defosse got left behind in Antwerp."

"Father told me."

"And I go over to Mr. Dunn in Dunkery twice a week for mathematics. Your father decided it wasn't his subject."

"I know."

The sweet smell of last year's apples seeped through the floorboards from the storeroom below.

She tried again. "Do you know why German kultur is spelt with a 'k'?"

He shook his head.

"Because the British command the 'c's." When he said nothing she said, "It's a joke, Gabriel, don't you understand? Command the seas – the navy, you know. Antony told it to me."

"I might have guessed. The Royal Naval Division doesn't seem to have grown him up much."

They sat in silence. Sunlight shone through the window on to the pale wood. The shadow of the frame cast black bars on the floor, marking out a noughts and crosses board at their feet. A pigeon scratched on the glass of the skylight and cooed.

"Frances says she isn't going to marry anyone, ever."

She meant, in some obscure way that was not clear, even to herself, to comfort him with the knowledge that he had no rival, but the moment she spoke she knew that not even that fact could console him.

He turned his head away. "She told me that, too."

Tentatively she touched his knee. "I am sorry, Gabriel."

"I know. So am I." He said, with an obvious effort, "Still . . . she wanted me to have one of her paintings. Where shall we start?"

She avoided his eyes as she brought the canvases over to the bed, frightened that he might still realise she had been fibbing, that it had never occurred to Frances to give him some remembrance of herself other than the letter so casually promised.

Gabriel's thoughts must have been far away. Surely he should have realised that none of the canvases were small enough to fit into a kitbag? But Sarah, glad to have a rare opportunity to look closely at Frances's work, did not tell him.

Even she, knowing little about art and pretending to know less, could tell that there was something special about her sister's pictures, the way the shapes, the intensity of light burst out of the flat surface of the canvas, the way in which the warm glow of evening light, cool mist of morning, the impression of rain held in the sky, was suggested by nothing more than a trace of colour, a touch of the brush. She looked at the paintings one by one, finished and unfinished, of Hillcrest, Somerset lanes and hills, the magnolia tree outside the window painted a hundred times over in different lights and different seasons, and remembered with shame her own fumbling attempt in Bristol. Did Frances realise how lucky she was to be able to catch for ever what would otherwise be only a fading memory?

Gabriel made no comment until he came to the last canvas when he said, "They're all too big, of course. I really want something small that would go in my pocket. Where does she keep her drawings?"

"In the chest over there."

He pulled out the shallow top drawer, brought it over to the bed and went slowly through the pieces of paper. "There seem to be a great many of you."

"She's always making me pose," Sarah complained. "It's not fair. Gwen and Julia refuse but I'm not allowed to say no because I'm the youngest. She gets cross, too, if you move. Julia doesn't mind but Frances won't even let you scratch your nose."

He pulled down his mouth. "Poor mouse. What a dreadful life you lead."

"It's not so bad really," she said sheepishly. "She does let me read."

He held up a watercolour sketch that Frances had painted last summer, of Sarah lying in the speckled light and shade of the verandah. Sarah thought for a moment, hoped, that that was to be his choice. She tried not to show her disappointment when he put it back in the drawer.

"She's a clever girl, your sister. More's the pity."

"She does like you, Gabriel, really she does. I know she does."

"Yes, of course. Liking's not enough, though, is it?"

How much was enough, she wondered, watching him replace the top drawer and take out the next. Liking – love – what were they? What was the difference between them?

All sorts of unpleasant thoughts came to mind as she watched Gabriel leaf through the drawings; thoughts that had never occurred to her when she jumped so impulsively from the scullery. Frances could be surprisingly methodical. Did she know what she had painted, keep lists perhaps? It was no good Sarah telling herself that Frances would have given Gabriel a painting had she thought of it. Of course she would, but Sarah offering one was a different matter. Frances, angry, acting on impulse, could be very frightening, could say things that hurt. It's for England, Sarah told herself, but there was a nasty hollow feeling in her stomach all the same.

He chose a self-portrait finally, a small drawing, little more than a preliminary sketch for something immediately abandoned, that caught in a few sepia lines, touched with white, the very essence of Frances, the liveliness of her spirit, the impudent attraction of her personality.

Sarah allowed herself a quiet sigh of relief. The drawing came from the back of the drawer, was old and creased. With luck, Frances would never notice that it was missing. Perhaps she had forgotten all about it.

"Do you know where she's gone?" Gabriel said, not looking at Sarah, as he put the sketch away in his wallet.

"She only told Annie."

"I know, but I thought you might know who she's with."

She shook her head.

"Or what part of the country. It isn't Dorset, is it?"

"I'm sorry, Gabriel. I don't know. I'd tell you if I did, truly I would."

"Yes, I believe you would. Well, I can hardly go round the country checking up on all her friends, can I? I suppose she realised that."

He stared bleakly out of the window while Sarah wished hopelessly that she could think of some way to make him happy.

"Will you write to me, when I'm in France?"

She was taken aback. "I wouldn't know what to say."

"You needn't say much — just little things so that I can think of you all at Hillcrest. Tell me what you're doing in your lessons, what Gwen's busy with in the garden. What Julia . . . and . . . well, you know the sort of thing. Don't look at me like that, mouse, I can't bear it."

She said slowly, "All right. I'll try. But if I write to you, you must write to me."

"That's blackmail."

She was indignant. "It's not. It wouldn't be fair, would it — me to write and you not?"

"The logical mind. All right. I'll write as soon as I get to France. Will you promise to reply?"

"I promise."

"Shake hands on it."

She felt very grown-up. His hand enclosed hers, pressing her fingers together until they hurt.

"Leave me alone now, there's a good girl. I'll stay here for a while on my own."

She hesitated in the doorway, suddenly remembering. "You will leave everything just as it was, won't you? So no-one knows . . ."

He turned his head sharply. She felt the hot colour rush up into her face. "You know what Frances is like. She can be awfully fussy sometimes."

"Yes, of course. Don't worry, I'll see to it. Off you go."

She could not believe that Gabriel would take the trouble to write to her. Yet if he wanted to know...

She did little writing these days. For some reason school stories had lost their fun now that she knew what school was like. She would enjoy writing to Gabriel.

Until last year she had never received letters. Now she never knew when there might not be one waiting for her on the mat – from Aunt Bessie, Geoffrey occasionally and, of course, Antony. Every few days there were letters from Antony. Although the youngest and the last Mackenzie to enlist he had been the first to go overseas, seen off by the King no less. At this moment he was cruising round the far end of the Mediterranean, sending back letters full of Homeric allusion in an attempt to beat the censor; unable, as the Rector wryly remarked, to decide whether he was taking part in a modern-day crusade to free Byzantium from the Turks or a re-enactment of the Greek expedition to Troy.

'I can scarcely believe my luck,' he wrote. 'Suppose I'd been born a month or two later – I'd have to go through the rest of life knowing I'd missed this by so little! Imagine it – soon I shall be walking through the streets of Constantinople!'

He never did. It came, that last letter, two days before the telegram announcing his death on a beach of the Dardanelles.

1916

Chapter Sixteen

Gwen glanced sideways at her eldest sister as she walked with Frances and Sarah over to the Rectory.

"Please try to behave tonight, Frances. It's very kind of Mrs. Mackenzie to have us. I'm sure I'd want to keep Gabriel to myself the first evening if I were her."

"I don't know what you mean," Frances said. "Of course I'll behave. I always do." Her eyes danced. "I tell you what — I won't say a word all evening."

"They'll think you're ill then," Gwen said. "Don't make the evening more difficult for them, that's all. I'm afraid Gabriel's coming home has brought up the misery of Antony all over again. I really did think they were beginning to get over it at last, but I don't know — Mrs. Mackenzie was very tearful this morning."

Sarah scarcely heard her sisters. The dull ache that had been with her ever since Antony's death the year before began to disappear, excitement to bubble up at the thought of seeing Gabriel again. Their relationship would be different now. After eighteen months of letter-writing, Gabriel would no longer look down on her, however affectionately, as his little mouse: they would be friends, proper friends.

It was difficult, looking back, to remember the agony those

first letters had caused her, the nail-biting misery as she tried to think what to say and how to say it. Only because she wanted his letters in reply did she persevere. Over the weeks writing became less of a burden until one day she realised that it had become one of her greatest pleasures, that the delight of telling him about them gave a gloss to even quite trivial incidents in her life. The original reason for the correspondence had long since been forgotten, weeks frequently passing with little or no mention of Frances. If Gabriel noticed or minded he never said so. Perhaps it no longer mattered when he and Frances corresponded too.

Sometimes Sarah wondered what went on between her sister and Gabriel. Had he forgiven her for her behaviour during his last leave? He never referred to it in letters to Sarah and Frances, who always kept her thoughts to herself, never confided. Sarah had no means of knowing.

The first sight of Gabriel shocked her. She had forgotten how much older he was, how large, how ... male. How could she have expected ...? Disappointed, suddenly shy, she could not find the words to join in the welcome.

There was a momentary awkwardness, a hesitation, at the dinner-table, remembering Antony, but Mr. and Mrs. Mackenzie were determined nothing should spoil Gabriel's homecoming. It was not as if Antony were the only member of the families missing, for both Geoffrey and Julia were in France.

"So Julia lied about her age, too," Gabriel said. "I wouldn't have thought it of her!"

"Sad though it might seem, lies are sometimes necessary," said Mrs. Mackenzie. It was Julia who had lied, not Frances, and in a worthy cause: Mrs. Mackenzie would not condemn.

"Oh, but I am shocked," Gabriel said, pulling down his mouth in mock disapproval. "Sarah would never do a thing like that – or would you, mouse? If it meant coming across to France to look after me you'd tell a hundred thousand fibs, wouldn't you?"

"I don't know," she said, and blushed.

"Frances might not, but you would. I know you. You wouldn't say anything but you'd come. Aren't you going to say anything tonight either?"

She searched for some subject of conversation. "Have you seen a tank?"

He shook his head. "I'm told they're fifty feet high and make so much noise that the Kaiser can hear them in Berlin."

He teased her through dinner, provoking her, drawing her out until her shyness disappeared and she was talking as much as — more than — anyone else, until she feared that Frances would reprimand her for being too forward.

But Frances smiled over the silver bowls of roses and said nothing. She sat beside Gabriel, as silent as she had promised, but radiant, vital, touching them all with her happiness. When Gabriel looked at Frances, Sarah knew that he had not merely forgiven her for her past behaviour but forgotten it as well.

After dinner Gabriel's medal was brought out, handed round and admired.

"I can't understand why he had it sent through the post," Mrs. Mackenzie said, "when he could have received it from the King himself."

"You didn't seriously think I'd waste a day in London when I could be down here, did you, Mother? Do put it away, everyone's seen it."

"Quite the little hero, aren't we?" Frances said. It was her first misdemeanour of the evening and deliberately done; she looked over to Mrs. Mackenzie as she spoke.

Mrs. Mackenzie sucked in her breath. "I'm surprised at you, Frances. It's because of men like Gabriel that we can sleep in our beds at night."

"What have I said?" Frances asked, opening her eyes very wide. "I agree with you. I realise only too well that if it weren't for Gabriel I'd have been raped in the stables by some nasty Hun long ago. In fact . . ."

"That's enough, Frances," Gabriel said severely, but his mouth twitched.

"I've been waiting in a fever of anticipation to hear all about the gallant deed," Frances finished.

"I'm sorry to disappoint you. I don't want to talk about it. It was over long ago – I've forgotten all about it."

Frances tilted her head up at him and laughed. Sarah was bitterly disappointed. Gabriel had said nothing in his letters to her about the incident that had won him the Military Cross and she had hoped that tonight she would hear every exciting detail from his own lips.

He refused to talk of battle or exploits, however, saying only that he had been glad to leave Flanders until he discovered the Somme to be worse, and suggested – with an ironic smile – that Frances should contribute to the war effort.

"Why not become a war artist? Think of it – sketched by Frances when going over the top, nursed by Julia when wounded! We might try it for a recruiting poster."

He sat in his favourite chair, smiling across the drawing-room at Frances, while Mrs. Mackenzie recounted every detail of parish news and Lucy played the latest tunes softly on the piano; until the evening light faded into darkness and Mrs. Mackenzie rang for the lamps to be brought in.

The squeak of the churchyard gate and the clang of iron on iron woke Sarah next morning. She lay in bed, trying to guess the time from the strength of light round the shutters, and listened for further sounds.

Birds sang. A peacock screeched from the Manor garden. There were no footsteps. Ghosts were silent but ghosts disappeared before dawn. There must be someone standing...

She tiptoed across the cold linoleum to the window and peered between the crack. Gabriel sat on the topmost step, leaning back against the wooden gate. He was wearing walking boots; his rucksack sat on the step beside him. Sarah lifted the bar from its socket, opened the shutters and was about to bang on the window when Frances came out of Hillcrest.

They stood in the middle of the road talking, while Sarah shivered in the grey light behind her window. They seemed in no hurry. After a while they went over and sat on the bottom of the churchyard steps for a long time in silence. Frances put her arm round Gabriel's shoulder, he leant his head against hers. There was a queer, uncomfortable pain in Sarah's chest. She could no longer bear to watch. All warmth had gone from her bed; she was still cold when Annie brought in her hot water and told her that it was time to get up.

Only two places were laid for breakfast.

"Frances has gone up to the Quantocks with Gabriel," Gwen said. "He needs to walk, she said."

Next morning when Sarah heard the clang of the gate latch she stayed in bed, and tried not to think of Frances and Gabriel with their arms round each other sitting on the churchyard step.

"Mr. Gabriel spends more time here than he does at the Rectory," Annie said. "I'm surprised your mother hasn't had something to say."

"She would like him to go over to Southway," Lucy said. "Miss Tuck's just taken them a telegram. Young Billy's been wounded. Mother thought it might help if Gabriel had a word with Mrs. Graham."

"He's up in the stables, I expect," Gwen said. "Sarah'll fetch him for you."

"Who is it?" Frances said when Sarah banged on the door at the top of the stairs. "Oh, it's you, Sarah. What do you want?"

Sarah stood in the doorway. Frances was on her stool, silhouetted against the window. Gabriel lay on the couch, propped up on one arm, the outline of his body on the silk bedspread repeated in the drawing on the easel. She hesitated.

There was something in the atmosphere ... she couldn't put a name to it. Happiness, warmth, contentment, love; all those and more, wrapped up and contained in the one room. But she was not part of it. She was outside looking in, separate, lonely.

"What do you want?" Frances asked impatiently.

"Lucy's here. Mrs. Mackenzie wants Gabriel to go down to Southway."

Gabriel looked up at Frances. "Shall I go?"

"It's up to you," Frances said and laughed. She sat at the easel but her brushes were still in the jug and the paint on her palette untouched.

"Tell Lucy I'm not here."

"But, Gabriel . . ."

"Or you could say I've gone to Dunkery for the afternoon."

She looked at Frances. "What shall I do?"

"You heard what he said. Off you go."

She went, slowly, reluctantly, down the path to where Lucy waited by the fig tree.

"Did you ask him?"

Her cheeks burned. "I don't know where he is."

"I thought Gwen said . . ." Lucy looked at Sarah's face. "Oh. Oh, I see. It doesn't matter. Don't worry, Sarah. I'll tell him at tea."

Annie sat on a deck-chair in the yard, reading *The Girls at his Billet*. Sarah crouched on the cobbles by her side.

"What's love like, Annie?"

"You'll find out in your own good time."

"Have you ever been in love?"

"Maybe I have, maybe I haven't."

"Do you always get married? When you're in love, I mean."

"That depends. Leave me be, there's a good girl. I've got to an exciting bit."

Sarah avoided Gabriel after the incident in the studio, not liking to tell lies, even on his behalf, and upset that he should expect her to do so. She retreated with her books to the lower lawn and it was while she was there on the last afternoon of Gabriel's leave that she became aware of voices. Absorbed in her reading, she had not realised at first that she was no longer alone.

Panic seized her. The long fronds of the willow under which she lay gave little cover, yet if she stood up to go Gabriel would surely see her and think that she had been eavesdropping. She could only lie still, pretend not to be there, not listen . . .

"Frances, do be sensible. Stop your drawing and pay attention. I'm serious. It may not be a fortune, two hundred and fifty a year, but it would make a lot of difference to you, wouldn't it? The war's bound to end sometime; it would mean you could go abroad occasionally. I know you're happy working here, but you've got to get away sometimes if your work's going to develop. You used to talk about Northern France or even Provence: two hundred and fifty would make it possible. Six weeks, a couple of months each year — there'd still be money over. And look at it from my point of view. If I knew I'd been of some use I don't think I'd mind so much. If I could think of you going on painting afterwards . . ."

"Dear me. I thought it was England that made dying worthwhile."

There was a long silence. He said, "You can be very cruel."

"All right, we'll discuss it sensibly. I'm to marry you for the sake of a widow's pension and my work. That's what you're suggesting, isn't it?"

"Well . . . I wouldn't put it quite like that."

"How would you put it then? Perhaps you'd care to explain what happens if you don't get killed. I'm stuck, aren't I? Married. What about my work then?"

"It's not much of a risk to take. No risk at all, I'd have thought, the way things are. Good God, Frances, wouldn't you be prepared . . . is the thought of me as a husband so repulsive?"

She ignored the question. "I want to get this absolutely

straight. You're suggesting that I should marry you in order to get your pension when you're dead. For money, in other words."

"I'm not suggesting anything of the sort. Of course I know you wouldn't do any such thing. I want you to marry me, I always have. I'd do anything to persuade you. And you love me, Frances, you know you do. It's only your work that's holding you back. I thought that if you could see it might help your work ..."

Her voice rose. "I don't think I've ever been so insulted in my life – you, of all people, to think I'd marry you, go to you, for money. Like a tart, like a common ..."

"I might have known you'd see it that way." His voice rose. "You never think of anyone else but yourself, do you? How do you think I feel? You haven't thought of that, have you? You don't care anyway, you've never cared. You're the only person that matters, sitting here in a little world of your own making, like a tin god ..."

"Oh yes. You can talk about tin gods. What about you at the Rectory? Now that Antony's gone you're the only one that matters – Gabriel says, Gabriel thinks, Gabriel wants. All day long. No-one gives a damn about Geoffrey or Lucy."

"Don't be so absurd. Lucy ..."

"Is taken for granted. Don't you contradict me. No-one wonders what Lucy thinks or wants. No-one cares. No-one has ever done a thing for her. Think back if you don't believe me. All those tennis parties before the war – those girls paraded in front of you for you to choose from. Why didn't anyone bring in a few men for Lucy, tell me that."

"You're being ... You know perfectly well that if it hadn't been for the war Lucy would have been sent out to my aunt in India."

"Oh yes. India – *India* of all places! What if she had found a husband out there? She's never been in India in all her life: she'd have been miserable. No-one thought to ask her if she wanted to go, did they, and she was much too nice to say what she thought. She ought to have married a curate years ago and finished up in a country parish. I wouldn't have thought that would have been too difficult for your mother to arrange."

"Why didn't you do something yourself then if you're so clever? I don't know how you can talk like that, Frances. You haven't done much better yourself, have you – refusing to send that child to school just because you made such a mess of school yourself."

"If you're going to start on Sarah . . ."

She nearly jumped up at that. No, no, no. You mustn't quarrel about me. Please don't quarrel about me. She buried her face in the grass, covered her ears with her hands. I can't bear it.

When she took her hands away Gabriel was saying in a voice she'd never heard him use before ". . . father to tan your backside. That would have sorted you out quickly enough. I've a good mind to do it myself."

"Get out. Get out of here this minute. Don't you dare set foot inside Hillcrest again. Ever, do you hear? Never, ever. Go on, get out."

"I'm going. You can have your little dictatorship. I hope it makes you happy. Thank goodness Julia had some sense. At least she managed to escape."

"I never want to see you again," she shouted and as he went up the slope threw her drawing board at him with all her strength. He never looked back. The board fell onto the grass with a dull thump and turned slowly over and over until it came to rest, drawing side down, on the flower bed. The thin red petals of a flattened dahlia peered out from under the wood like fingers, pointing, accusing.

Ssk, ssk, went Annie's chair, rocking gently to and fro. *Ssk, ssk*.

"Could do with young Jess now, to turn the heel. Never was much good at knitting. Drink up, Miss Sarah, afore it goes cold." She glanced up from the needles. Her voice changed. "Oh, love, what is it? What's the matter?"

Sarah wiped her nose on her sleeve. "Nothing."

"You don't cry about nothing. Mr. Gabriel going back – is that it? Oh love, don't take on so. He knows how to look after himself, Mr. Gabriel does. It doesn't do to worry."

Sarah shook her head. "It's not that."

"What then? I can't help if I don't know."

Sarah's voice shook. "Why do people have to quarrel?"

"Oh, that's it, is it?" Annie said in quite a different voice. She put down her knitting and looked at Sarah with worried eyes. "Everybody gets cross now and then," she said at last. "A bit edgy sometimes – you know that. Take Miss Frances and me. We have our ups and our downs. They don't mean anything. I'd give my right arm for Miss Frances and I daresay she'd do the same for me." She thought for a moment. "Well, her left arm anyway."

"Mr. and Mrs. Mackenzie don't quarrel."

"You don't know what goes on in people's homes when you're not there to hear. I daresay you're right about Mr. and Mrs. Mackenzie. It doesn't mean they think any better of each other. Quarrelling doesn't mean you don't care, don't think that."

"It wasn't not caring. If you'd heard them, Annie. They hated each other, really they did. Frances said ... dreadful things ... and he's going away tomorrow. How could she?"

"Ah, Miss Frances. Miss Frances isn't like you and me, love. She's a very clever girl. Them who know about such things think very highly of her work, so I'm told. Clever people can be a bit difficult to live with, you know. It's just one of those things."

Sarah was indignant. "Gabriel isn't difficult."

"Nobody's perfect," Annie said tartly, "except the good Lord and I daresay He tried His mother sorely when He was a child. Mr. Gabriel has his moments, I shouldn't wonder." She picked up her knitting. "Time you were in bed. You don't want to show Mr. Gabriel a face all red-eyed and weepy when you see him off tomorrow. Go and say goodnight to your sisters now."

Gwen was frowning over her watercolours in the fading light of the living-room. A single rose curved over the edge of the specimen vase towards the table.

"Julia'd say you'll ruin your eyes," Sarah said from the door.

"Just as well she's not here then. If I don't finish tonight it'll be too late, that's the trouble. The petals will have fallen. Off to bed?"

"I suppose so. Where's Frances?"

"Don't bother to wait up for her. She's gone out. Sleep well."

Sarah stood at her bedroom window, trying to understand. Such bitterness, and yet the other afternoon in the studio ... and that first evening at the Rectory ...

The leaves of the chestnut trees were beginning to fall. In a week or two Gabriel's window would be visible through the bare branches but tonight there was no way of knowing whether he was already in his room, thinking of Frances, or still in the drawing-room trying to forget.

The Mackenzies always refused to be seen off at the station. Farewells were said at the Rectory gate. Not this time. This time Frances went into Taunton with Gabriel in the taxi.

Only Sarah saw her on her return, stumbling past the window of the storeroom where Sarah was putting apples away for winter. Sarah carried on automatically, wrapping each apple in its piece of oiled paper, laying it carefully on the slatted shelf, while she thought about Frances. Brilliant, sparkling Frances whom people turned to look at and admire; Frances swollen-faced, red-eyed and ugly, crying alone in her stable studio.

1917

Chapter Seventeen

Sarah was growing up; an experience which was not altogether what she had expected. She had looked forward to staying up late, doing what she wanted, being able to say no if she wished. She had not thought about the onset of responsibility, or growth of a conscience. She found it difficult now to lie unheeding with a book when others were calling her; impossible to say with her previous wide-eyed innocence, "But I didn't hear you," when it was only too obvious that she had.

She had responsibilities beyond the ordinary household tasks and her schoolwork. The hens were in her charge, the eggs they laid her contribution to the war effort just as the vegetables that went from Hillcrest to the Red Cross Hospital in Taunton every Friday were Gwen's. Sixty thousand eggs a week the base hospitals in France required. Among that number the few dozen that Sarah supplied were but a drop in the ocean, yet she hoped every evening as she gently wiped off the earth and removed the feathers still clinging to the shells that a Hillcrest egg might one day reach Julia's plate in Étaples.

Because she was paid for all the eggs that were taken, she learnt to keep accounts. Money received, money spent on meal for the hens, the number of eggs sold: once a week she sat at the kitchen table with Annie and Gwen and struggled with figures until her head ached.

The egg money went into the household purse. What she earned spudding was her own, to spend as she wished. Not that there was much to do with it. She had hoped to go into Taunton to see *The Battle of the Ancre and the Advance of the Tanks* at the Lyceum, but Frances said no.

"It's really happening, Sarah. Real men, being killed. You don't want to see that. The whole idea's macabre. And don't ask Annie to take you, not with Dick at the front and Jim a prisoner in Germany."

It was not the battle that Sarah was anxious to see so much as the tanks, but she knew that she would never be allowed into Taunton on her own. Nor was there any point in asking Gwen to accompany her. Not even the prospect of seeing a tank would lure Gwen out of the garden in the sowing season.

Sarah went spudding for Mr. Escott who paid her a shilling a day to weed Tinker's Meadow, the field beyond the elms. In time she came to feel an almost possessive pride in the field and would often slip through the gate by the stables to admire the growing crop stretching away to the distant hedge unblemished by rusty spike of dock or purple thistle head. She began to understand why she sometimes found Gwen surveying the vegetables, silent and immobile for minutes at a time, with a smile of quiet satisfaction on her lips.

She spent more time with Gwen in the garden these days, not because she was asked or told to help but because she enjoyed it. She and Gwen were growing closer. In the garden Gwen discussed things with her, asked her opinion, talked to her as an equal. Gwen had always been quieter than Frances and Julia and more contained, just as her paintings and drawings were less flamboyant and more introspective. It surprised Sarah to discover some of Gwen's ideas and opinions. She looked with different eyes at people as a result, wondered what was going on in their minds, what they were thinking.

For the first time she began to sense that she might not always lag behind her sisters, that some time she would catch them up; would eventually reach a stage when the years between them no longer mattered. It was only dimly realised; she could not have put the thought into words. And for most of the time – playing with the public schoolboys who lived under canvas in the Manor grounds every August while helping with the harvest; getting her shoes wet fishing for sticklebacks in the stream by Nanny Mackenzie's – she was a child still. Yet deep inside her, even in the grey days of a dismal winter, there was a warmth, unexplained and barely understood, like a promise for the future.

She disliked the winter, the dripping trees and undergrowth, the white mist that filled the lanes round the village. She hated the dampness that seeped into the house, the colds and chilblains that winter brought, but more than anything else she disliked having to stay indoors where every action was noticed and commented on. "You're not scribbling *again* . . ." Why should they say that when they were always drawing themselves? "You'll ruin your eyesight in this light." So equally well might they. They meant well, she knew. It never occurred to them that she needed to be alone, hated to be watched.

But winter was passing. Narcissi had already replaced the snowdrops along the verandah edge. Soon it would be spring and then summer once more, a summer that would remain forever in Sarah's memory lit by a pure golden happiness that she would never know again; the last summer of her childhood.

It was Sir James who brought the news to the Rectory. Returning from Bristol where he had been visiting his sister's wounded son, he announced that he had caught sight of Gabriel's name among those listed on the door of an adjoining ward.

"I do think it was thoughtless of him not to find out more," Lucy said when she came over to tell the Purcells; it was the nearest they had ever heard her get to criticism. "It wouldn't have taken a minute to look round the door and say hello or ask the sister what was wrong. Sir James said he didn't want to miss his train, but trains go all the time."

"Do you think it really is Gabriel?" Gwen asked. "It seems odd that you haven't been told, if there's been time to bring him all the way from France."

"Sir James thinks it must be. He says the War Office gets behindhand if ... well, if there are a lot of telegrams to send. Gabriel and Father have their own private code, you know. Father thinks he's near Arras – was near Arras, I mean – and that's where the fighting has been. And we haven't heard from him for ages. You haven't either, have you, Frances?"

Perhaps it was guilty conscience that made Sir James offer to drive Mr. and Mrs. Mackenzie to Bristol for the day; perhaps it was only his usual kindliness. They would be able to visit Southmead Hospital, he said, and set their minds at rest.

The Purcells and Lucy stood on the Rectory steps to see them off. Mrs. Mackenzie was already in the Daimler and Mr. Mackenzie, sallow-faced with anxiety, about to join her when Miss Tuck came panting up the back way from the village, flat feet flapping awkwardly on the drive. Tears ran down her cheeks.

"Oh, Rector, sir ... Oh, Mrs. Mackenzie ... oh, sir."

Colour drained from Mrs. Mackenzie's face at the sight of the yellow envelope. "Geoffrey ..."

Except that it wasn't Geoffrey. It was Gabriel ... missing, believed killed.

When the smelling salts had been found and administered and Mrs. Mackenzie was recovering on the couch in the drawing-room, the Rector and Sir James discussed what to do.

"It can't be Gabriel," Mr. Mackenzie said. " 'Believed

killed' – the War Office wouldn't say that unless there was good reason for thinking it. You must have misread the name."

"There's nothing wrong with my eyesight," Sir James said. " 'Captain Gabriel Mackenzie' – it was quite clear. Of course it's your son. Mackenzies may be thick on the ground but how many are called after one of the heavenly host, tell me that. *And* the correct rank. With the number of telegrams going out these days it's not surprising if there's a mistake now and then. We'll go up to Bristol right away, the two of us. We'll telegraph back here as soon as we know the situation. The girls are here to look after your wife if the news is bad. But it won't be, I'm sure of that. Between ourselves," he lowered his voice, "it did occur to me last night that it might be a mistake to take Mrs. Mackenzie until you know how badly he's hurt. The lad next to my nephew's had his jaw shot away – not a pretty sight, I can tell you."

He had drawn the Rector out of Mrs. Mackenzie's hearing but Sarah heard and so did Frances. Frances looked like Sarah felt, slightly sick and frozen so cold and stiff that it was difficult to move or talk, difficult even to breathe.

They sat in the Rectory drawing-room after Mr. Mackenzie and Sir James had departed, avoiding each other's eyes, trying to make conversation, waiting. Sarah wondered whether to go into the study and get on with her work but was reluctant to leave. Frances had no such scruples.

"I can't stay here," she said. "I'll be in the stables if you want me. Perhaps someone would let me know ..."

There had been snow at Easter but today was warmer. Bertha had lifted the window a little; the scent of the narcissi outside floated in under the sash.

It was not until the afternoon that the telegram from Sir James arrived. The doorbell rang; there was a mumble of voices; footsteps crossing the hall. The salver shook in Hilda's usually steady hand. Unable to watch, Sarah looked out of the window at Shattock planting dahlias in the garden. Stakes marked the holes already dug for the tubers waiting on the grass.

"Oh heavens," Lucy said, "where did we put the smelling salts?"

Mrs. Mackenzie lay back on the couch, the paper resting where it had fallen, on the floor.

"Come on, Mother. Breathe in. It's all right. Deep breaths. Sarah, you'd better go and tell Frances."

Sarah picked up the telegram, afraid to read it; when she had done so, unable to believe the words. Then she was running back to Hillcrest, running up the road, stumbling over the rutted tracks in Tinker's Lane. There was a pain in her side. Her throat hurt. She started calling before she reached the back gate.

"Frances, it's all right. The telegram's come. It is Gabriel. He's all right. Frances!"

Frances's face appeared momentarily at the studio window. She half jumped, half fell down the stairs, nearly knocking Sarah over at the bottom. Her face was pale green, almost translucent, like bone china held up to the light. She clutched Sarah.

"Are you sure?"

"Of course I'm sure. You're not going to faint, are you? Mrs. Mackenzie's fainted twice today."

"Of course not." She gave a laugh that was almost a sob, holding on to Sarah as if needing support. "What did it say? Tell me, tell me!"

"It's his leg. Serious, it said. What do you think that means? He sends everyone his love. I expect that means you really. Oh, Frances, isn't it wonderful? And Bristol, too. You'll be able to go up and see him."

As it turned out there was no need to travel to Bristol – Sir James arranged for Gabriel to be brought home to the Rectory. Mrs. Mackenzie was not inexperienced; Sister Vicary, an army sister now retired, would be able to help with the dressings; Dr. Milne was used to injuries to limbs among the farming community in which he worked; and if extra help were needed Nanny Mackenzie could be brought out of retirement.

"When you think about it," said Sir James, "we're doing the hospital a favour, releasing the bed for some other poor fellow."

Mr. and Mrs. Mackenzie said little. When the War Office forbade wounded officers even to convalesce with their families, preferring to send them instead to army convalescent homes in the hope that early nights and a quiet life would encourage a quick return to the front, the Rector and his wife were reluctant to enquire too closely into how their particular miracle had been performed. Sir James merely remarked that anything was possible if you knew the right people. A Justice of the Peace himself, and well known in the county, he presumably did know the right people, or at least those who mattered when it came to spiriting a disabled captain out of Southmead Hospital.

As soon as Lucy brought word, the Purcells went over to the Rectory to visit. Mrs. Mackenzie stood on guard by Gabriel's pillow.

"Not merely a hero this time, but a wounded hero," he said to Frances. The mocking tones sounded forced. His face looked all bone and purple shadows. "I trust you're impressed."

Frances was subdued, tense. "I suppose you were trying to get a bar for that stupid cross of yours."

The wry smile brought the old Gabriel to life. "Alas, the part I played this time was totally undistinguished. Went down in the first wave, in fact."

"I don't understand," Gwen said. "The War Office didn't just say you were missing. They said you were believed killed. And the telegram came ages after it must have happened, after Sir James had found you in that hospital."

"Not the most efficient of organisations, the War Office,"

[174]

Gabriel said. "I suppose someone must have seen me go down and took it for granted I was dead. It can't have occurred to anyone, when the body didn't turn up, that it was already on its way to Blighty." He gave a weak smile. "I thought I'd been killed myself, as a matter of fact. It was quite a shock to find that I was still alive. If it hadn't been for my servant hauling me back to the main dressing station ... He'd been badly hurt in the arm. I couldn't walk and he wasn't much good at carrying – we must have been a comic sight, when you think about it."

Sarah would have preferred not to think about it. She stood by the window, not knowing what to say, upset by his pallor, the limp way in which he lay against the pillows. She tried not to look at the mound in the bed made by the cradle under the blankets but against her will her glance kept returning.

"The papers said it was a splendid victory," she said.

"You don't want to believe everything you read in the papers. What have you got there, mouse?"

"Annie made you some calves' foot jelly." She said defensively. "It's supposed to be good for you. She went to a lot of trouble making it."

"She doesn't seriously think I'm going to eat calves' foot jelly! Someone – not you, Frances, you don't know how to say no gracefully – someone had better tell Annie that I am not ill, I have merely been wounded. If she wants to make me anything she can bake me one of her fruit cakes. The biggest possible."

"But Gabriel . . ." Sarah said and stopped, looking round at the others. Had he not heard about food shortages, did he not know of the submarines in the Atlantic?

Mrs. Mackenzie glanced at her watch. "He found the journey very tiring. I think ..."

Gabriel caught Frances's hand. "Not you," he said. "Please stay."

Sarah walked silently back with Gwen to Hillcrest trying

[175]

not to think of the hummock in the bed, nor the wounded leg that lay beneath it.

Such thoughts did not trouble Frances. She was singing when she returned later from the Rectory. Sarah, washing the last of the dried fruit in the kitchen with Annie, was shocked. "She sounds so ... so happy."

"Of course she is. So should you be. It's a nasty do, I daresay, but it might be a lot worse. If Mr. Gabriel's here in his bed he's not going to get killed in France, is he? Bertha told me he could be here for weeks. Maybe there'll be peace by the time he's ready to go back. Of course Miss Frances is pleased."

Chapter Eighteen

It was Mr. Mackenzie who decided that Gabriel should take over Sarah's lessons, thinking that the company and attention required would take Gabriel's mind off the pain in his leg. It never occurred to Mr. Mackenzie that Sarah would be anything but pleased and he would have been distressed had he realised with what apprehension and reluctance she crept up the Rectory stairs that first morning. She found it difficult to come to terms with the sight of the usually active Gabriel lying helpless in bed. The man to whom she wrote in France and the wounded soldier here in the Rectory bedroom were two separate people, neither of whom she knew.

"So," Gabriel said when she slipped nervously round the door, "we're giving Father a rest. Have you brought up your work? Sit down, while I see what you've been doing."

In Gabriel's room there was not the litter of books and papers, scattered where left and not put away, that Annie so deplored in Sarah's own bedroom: Mrs. Mackenzie believed that sickrooms should contain nothing but the necessities of nursing. Gabriel's desk, closed and unadorned, stood in one corner; a bare-topped table under the window. Even his bedside table had only a clock and a glass of water on it. Mrs. Mackenzie had stopped short at clearing the wall, however.

Photographs of schoolboy and university teams still decorated one wall; on the wall facing the bed hung an oil painting of Frances.

Julia was the portrait painter in Hillcrest. When Frances painted someone she painted him as part of the whole, made a picture of sitter and background. Julia went deeper into the character, showed the sitter as an individual.

"Warts and all," Antony had said of Julia's portrait of Lady Donne and laughed. Sarah had not understood: Lady Donne had no warts that Sarah could see and Julia had shown none. Lady Donne herself had disliked the portrait, saying that Julia had made her look petulant when she should have been gracious.

Perhaps Julia should have done the same with her portrait of Frances. Frances's head was tilted back, eyes glinting under half open lids, teeth catching her lower lip in the beginning of a smile. It was a typical pose – Frances about to make an outrageous statement in order to provoke, to annoy. Sarah sat tautly in her chair, hands clutching each other in her lap, and wondered what Mrs. Mackenzie thought as she fussed about the room watched by Frances's mocking gaze.

The bustle of an everyday Rectory morning went on outside the bedroom. The new housemaid was banging the paintwork not far away. At any moment Mrs. Mackenzie would come upstairs to reprimand her. Perhaps Mrs. Mackenzie would come in to see how Gabriel was and decide that he was not well enough to give Sarah lessons.

"Well, well," Gabriel said at last, smiling across at Sarah. "Quite bright, aren't you? You must be well ahead with your work. When does Father expect you to matriculate?"

"I don't know."

He gave her a sharp look. "You're not frightened of me, are you?"

She shook her head.

"We know each other well enough after all those letters, surely?"

"Yes."

Mr. Mackenzie expected her to make mistakes from time to time, to have occasional difficulties in understanding. He was patient in explaining and correcting. She was not sure what Gabriel would expect. Gabriel was clever; a Fellow, whatever that might be, at Cambridge. She could not bear to be thought stupid by Gabriel.

"Perhaps we should just talk today. My leg's not very comfortable; I don't think I'd be much good at teaching."

On the third morning it was eleven o'clock before she reached Gabriel's bedroom.

He laid down the *New Statesman*. "I thought you'd taken fright and run off."

"I have a French lesson with Madame Defosse on Thursdays."

"Yes. Mother remembered to tell me eventually. What have you learnt today?"

"I didn't learn anything — it was literature and elocution." She made a face. "Madame Defosse is dreadfully fussy about elocution. I have to learn something by heart and then I say it aloud to her and she corrects my pronunciation." She rattled off very fast with exaggerated gestures:

> *"Maître Corbeau sur un arbre perché,*
> *Tenait en son bec un fromage.*
> *Maître Renard, par l'odeur alléché,*
> *Lui tint à peu près ce langage."*

"That sounds splendid — Frencher than the French. What does it mean?"

She stared. "Didn't you understand?"

"Of course not. I don't know French. Oh, enough to get by with, over there, but German's my language. I thought you knew that. I had a German nurserymaid when I was a child and then a German governess — I spoke German better than English when I was small. That's the only reason I got the

[179]

M.C. really – the Huns would never have surrendered if I hadn't shouted at them in German. There were six of them in the trench; they could have finished me off without any trouble if they'd had their wits about them."

She could not believe that she knew more than Gabriel. "You really don't understand?"

"You'll have to teach me. Tell me what it means."

"It's a fable, by La Fontaine. He wrote lots of fables. They all have morals. This one's about flattery, being taken in by fine words. 'Maître Corbeau' – that's Master Crow ... "

She lost her fear. If Gabriel, grown-up, clever, could admit to ignorance, why should she mind making the occasional mistake? Within days the two Gabriels had merged into the person she had always known. She forgot about the wounded leg, never noticed the hummock in the bed, could hardly wait for the grandfather clock to strike nine o'clock and during the weekend waited impatiently for Monday morning.

Unlike Mr. Mackenzie, who was a quiet man, Gabriel encouraged Sarah to talk during lessons, often not about schoolwork at all but about the world outside the Rectory, outside the village; about growing up ...

"Do you think you could persuade Frances to let me put my hair up?" Sarah asked him one morning. "She says I'm not old enough but I shall be sixteen the year after next."

He shook his head. "I like you as you are. I shan't be able to call you my little mouse when you grow up."

She piled her hair on top of her head and regarded herself critically in the mirror. His reflection watched her from the bed. "You're improving, you know. Given time you might even grow into a beauty."

She turned to look at him. "Do you really think so? Do you think I'll ever look like Frances?"

His mouth quirked up on one side. "Why not try looking like yourself?"

She twisted her hair behind her neck and tried to imagine it cut. "There was a girl at school who had short hair."

"It's quite the rage in London, I believe."

"It must be easy to look after, don't you think?"

"Maybe. But be warned: if any Purcell shears off her crowning glory I shall refuse to step inside Hillcrest ever again."

He was joking of course. Or was he? It was sometimes difficult to tell whether Gabriel was teasing or not.

He was always serious about her work, however. Though he made a game of lessons he expected high standards and she had to strive to earn his praise.

She came dancing home one lunch-time to an interminable discussion between Gwen and Frances on the possibility of replacing unobtainable sugar with dates for making jam. When they got onto how it might affect the taste, Sarah could bear it no longer and burst into the conversation like a small explosion: "Gabriel says I write Greek prose with a great sense of style!" She could not understand why her sisters laughed.

Not that Gabriel often praised. He said what he thought and it was not always approving. Sarah had never realised before how strongly he felt about her going to school.

"I never heard anything so ridiculous in my life: an eleven-year-old saying she didn't want to go to school and *not being sent*! You can't always have what you want in life, or haven't you discovered that yet? It's supposed to be a very good school – what made you dislike it so?"

"I don't know. The other girls, I think."

"Why? Did they bully you?"

She remembered how she had been surrounded by them in the playground, trapped. Looking back, she thought now that perhaps they had been curious, interested, rather than threatening. And Kate had shown her round, even offered to keep her the desk next to hers. That remark about being 'pi' had probably not been unkindly meant.

"No," she said. "It was just that I wasn't like them. I was different."

"Of course you were. You'd never been to any school before, let alone that one. In a month or two you'd have been indistinguishable from the rest."

"But I was different, Gabriel. I'm an orphan."

"Oh, come on. I suppose it never occurred to you that there might be others there without parents?"

She shook her head.

"Well then. Just think of all the fatherless children there must be in the world today. Oh, Sarah, I know your parents are dead and I'm sorry for it, but you've got mine, you've got your sisters, Lucy, me – I'd have thought you were rather better off than a lot of children. You have two families instead of one."

She flushed. "That wasn't the only reason. There were others. The uniform, for instance. Everyone wore uniform."

"You'd have had the same."

"Would I? I thought I'd have Gwen's old uniform. It's been handed down twice already and anyway it wasn't the right colour."

"Oh, Sarah, really. No-one can read your mind – why didn't you tell Frances you thought that? Or Julia. She was still here then, wasn't she?"

"It isn't very easy to talk to Frances. Particularly about clothes. Ordinary clothes I mean, not uniform. She always knows best. She looks at you as if she thinks you're rather odd. As if you have peculiar ideas."

"There's an easy solution if you don't like the clothes Frances chooses for you. Why don't you make your own?"

"Make my own?"

"Why not? Frances has made hers ever since I've known her."

"I wouldn't know how."

"You're spoilt, that's your trouble. Ask Frances or Gwen; they'd be happy to help you get started."

She was upset. "You don't really think I'm spoilt, do you, Gabriel?"

"I think you take a lot for granted."

Later he said, "You know, mouse, Frances would be very hurt if she thought you couldn't discuss things with her."

"You haven't told her, have you? Oh, Gabriel, don't tell her, please don't. I thought it was private, what we talk about, I didn't think you'd tell anyone . . ."

"Hey, calm down. I won't repeat anything you don't want me to. You talk as if Frances is an ogre, that's all."

"Of course she's not. I can talk to her about some things. Going to school, for instance. She was very nice about that, she understood how I felt. I suppose it's clothes mostly. But, Gabriel, you don't know how nice it would be to have something smart to wear . . . something a bit grand . . ."

"Latin epigrams today, I understand," Frances said, returning from the Rectory one afternoon. "What it is to be bright – I don't even know what an epigram is."

"Oh, Frances!" Sarah thought she was joking. "The trouble is, Mr. Mackenzie thinks I'm doing lessons with Gabriel. But I'm not, a lot of the time I'm not. We just talk and have fun and play games. I don't know what Mr. Mackenzie would say if he knew."

Her sisters glanced at each other. "I shouldn't worry," Frances said. "You're probably learning without realising it. Gabriel must discuss what you're doing with Mr. Mackenzie."

Sarah was doubtful. There was always the thought of the Oxford entrance examination, rearing up like Mount Everest in the distant future.

"I told you you should have gone to that school," Gabriel said. "It must be easier to get into Oxford from there than from here."

She tried to sound casual. "I don't know that I want to go to Oxford."

"I take it that that's another subject you haven't discussed with Frances. Or my father, for that matter."

She looked sheepish.

"May I ask why not?"

"Well." She tried to organise her thoughts. "What good is it? I mean, if you spend three years at art school you can paint pictures and sell them and earn some money. Or exhibit them – did you know Frances's having an exhibition in London this summer? – and people sometimes buy those, too. But three years at university, all that time and money, and then what do you do at the end of it?"

"How about teaching? Teaching gives you a regular income, if that's what you're worried about, which is more than can be said for painting."

"I hadn't thought of that." She remembered the classics mistress in Bristol. "I think I might enjoy teaching."

"Or you could write."

"Books, do you mean?"

"Perhaps. Or short stories, articles. It's too early to tell. You couldn't expect to make a living out of it, certainly at first, but if you went in for teaching you'd have time to write as well. I'm sure you have definite talent in that direction, mouse. I really do look forward to your letters. They're so vivid; you have a nice turn of phrase and the dialogue's superb."

"Dialogue's easy though – all you're doing is repeating what people have said. It's description that I find so difficult. It's all so clear in my mind, you see. When I shut my eyes the pictures are there on the back of my eyelids but I can never find the right words . . . "

"That'll come with practice. Or perhaps you've made a mistake and should paint instead."

"You know I can't paint."

"I know Frances says you can't. I've sometimes wondered whether, with the age gap between you, she wasn't expecting the impossible. A ten-year-old to paint like a seventeen-year-old, for instance."

"We had an awful art lesson at that school. As a matter of

fact I didn't think much of the other girls' paintings. I could have done better myself."

"You see? You'd better get out your paintbrushes."

She shook her head. "It's too late now."

"Mind you, you're making pictures all the time when you write. Except that whereas the painter provides a picture already there, on the canvas, a writer provides the reader with the ingredients to make his own. More satisfying for the recipient, I think, though I'm not sure about the creator. Painters are generally much happier people than writers, it seems to me; they appear to get more satisfaction out of the actual act of creation."

"Frances is never satisfied."

"Not with the final result, no. Self-satisfaction is a sign of mediocrity. One should always feel that one could have done that little bit better, whatever the form of creation. Listen to Gwen talking about the garden – there's a dissatisfied artist if ever there was one! It must be difficult being a gardener, working on such a long time scale. Not days or weeks, but years."

She gave a great deal of thought to Gabriel's suggestions. The prospect of a definite future made her feel secure, more hopeful about going to Oxford; the possibility of a writing career gave purpose to her past scribblings. She sorted out her stories, some barely started, others abandoned in the middle, yet others that she could see now had been modelled too much on past favourites like Harrison Ainsworth or Conan Doyle. She chose one that she was reasonably pleased with and took it over to the Rectory for Gabriel to read, marvelling at her boldness, knowing that she would die rather than let anyone else see it.

"The writing's not bad at all," Gabriel said when he put it down. His mouth twitched. "Oh, Sarah, you'll be the death of me. Wherever did you get your ideas about the goings-on in art schools?"

"From Jess," she said, pink-cheeked.

"Jess Hancock? And where did she . . . ?"

"From Mary, I think."

"And I suppose Mary got it from Bill. Or more likely some trashy magazine. Did it never occur to you to talk to Frances?"

"She said all they did was work."

"You didn't think that she might have been right? Sarah, you don't *know* anything about art schools. You must write about things you know. You've only got to think of H. G. Wells to realise that, or the Brontës."

"How can I write about Huish Priory? Nothing ever happens here."

"All right. Think about Jane Austen. And anyway I don't know how you can say such a thing when you've been writing me letters full of incident for years. Use the setting you know. Invent the plot but put it in a village setting. That way you'll make it sound right. Do you understand? As for incidents, what about the business with Count Alpbach and Farrance House at the beginning of the war. Wouldn't that make a splendid climax to a story?"

"Yes. Yes, it would." She remembered Bert Hawkins coming up to Hillcrest afterwards, recounting the story over a mug of tea in the kitchen while she and Annie listened, open-mouthed. "What a good idea. I'll think about it."

"How is young Jess these days?"

"I don't know," Sarah said.

She had gone down to see Jess on Jess's first visit home, to offer her books to read.

"I don't get much time for that now," Jess had said, speaking coldly, looking at Sarah as if at a stranger. She had spent her holiday helping Mary Roberts with young Willy and had kept away from Hillcrest. Sarah had been hurt, though she said nothing, and in future years waited, but in vain, for Jess to make the first advance.

"I don't see much of her," Sarah said now. "She only comes home once a year."

"Poor Jess. If ever a child should have gone on with her education she should."

Sarah stared. "But Gabriel, how could the Hancocks have managed to send her to Bishop Fox's?"

"I know. That's the pity of it."

"Besides . . ." Sarah said and hesitated.

"Besides? If you couldn't cope with school Jess would be unlikely to either?"

"Yes," Sarah said. It was not what she had meant to say but she was not sure how to put her meaning into words.

When Sarah came to go home that morning Gabriel said, "You know, mouse, I'd keep off passion if I were you, until you're old enough to know something about it."

"But it'll be years and years . . . sometimes I think I'll never grow up."

He moved restlessly in bed. "I don't know why you should be so anxious. There's nothing wonderful about being grown-up, let me tell you. Enjoy yourself while you can."

He helped her frequently with her writing after that. He was a strict mentor, making her write and rewrite, forcing her to justify everything, even the choice of words. By now she was enough at ease with him to argue when she disagreed.

"You are an amazing child," he said once. "So quiet, yet such definite opinions!"

"I think you're wrong," she said stubbornly.

"Maybe I am. You must listen to what I say though and my reasons for saying it. Then, if when you've considered you still think I'm wrong you can stick to your own opinion. That goes for everything, not just writing. Work, politics, life in general." He smiled. "Your letters never cease to astonish me, you know. Father always talked about still waters but it never occurred to me that such deep thoughts went on behind that solemn countenance of yours. Once you'd relaxed that is, and got over the stiffness of those first letters, the yours faithfully bit. Why did you say that? Didn't you want to send me love?"

She blushed. "It seemed a bit . . . familiar."

He tried to persuade her to write poetry, telling her that, like Greek prose, it was excellent training in the use of language. She was not persuaded. She liked poetry for its sound rather than its meaning, she liked hearing it read or reciting it herself inside her head, but she hated writing it.

"Do you write poetry, Gabriel?"

"Very little now. Only to Frances."

But Frances doesn't know anything about poetry, Sarah wanted to say. Frances wouldn't understand it, wouldn't appreciate it.

"Your father told me once that Antony wrote poetry. Did you know that?"

"Yes. He used to show it to me occasionally and ask my advice."

"Was it any good?"

He was silent for a moment. "He was very young of course, but yes, I thought it showed great potential." He sighed. "Such a waste. I tell myself that even if he'd waited he'd have been dragged into it by now but it doesn't seem to help much, even knowing that."

A lump came into her throat at the thought that Antony's poems might well have been among those he had read to her in the orchard or under the medlar tree, that she had never realised or appreciated them.

She had never said Antony's name aloud since the day of his memorial service when she had walked home across the churchyard with the organ notes of the Dead March and the muffled peal of the church bells clanging together against the inside of her head. When the Rector mentioned his youngest son she was silent, as if putting into words would make real a fact that she was not yet prepared to accept. To her surprise she found now that she could talk to Gabriel about Antony and gain relief in doing so.

"You must miss him very much," Gabriel said.

She had sensed, even when Antony was alive, that he was growing away from her; that his interests were becoming adult while hers were still those of a child, but it was only Gabriel's company that made her see how lonely she had been after Antony's departure.

"It's all right now," she said.

It was more than all right. It was as if the happiness of those two days on the Quantocks, far back in the past but so well remembered, had been magnified and spread over the long days of summer.

Chapter Nineteen

Gabriel was soon absorbed into the Rectory routine, or perhaps, quietly and unsuspected, he reorganised the routine to suit himself. In the mornings he taught Sarah. In the afternoons he was entertained by Frances who came over as soon as Mr. and Mrs. Mackenzie had departed on the parish rounds.

Frances was easy to get on with these days, painting all hours of daylight, laughing, teasing Sarah, even charming Mrs. Mackenzie. "I can't have the old trout deciding her son's had a relapse and is too ill to see me," she said, adding, "Don't you ever let me hear *you* talk like that, Sarah!"

In time, of course, Gabriel was able to come downstairs. Sarah hated that. Up in his bedroom she had him to herself; downstairs she had to share him with his parents, the rest of the household, any parishioner who might call.

Once up, Gabriel wasted no time in borrowing a pair of crutches from the Red Cross and hobbling over to Hillcrest. The routine changed: he spent every afternoon with the Purcells, staying until the Rector called after evensong to help him back to the Rectory.

He needed no entertainment. He lay on a couch on the lawn, dozed, or read, was painted by Frances, talked to whoever wanted to converse. Knowing he was there, people dropped in – Sir James and Lady Donne, Mr. Tasker from Clay Court. Even Colonel Sherwood walked across the fields from Dunkery St. Michael to discuss the South African campaign and compare it with the campaign now being waged in France.

"It's like Madame de Pompadour," Frances said. "Wasn't she the one who held court and soirées and that sort of thing?"

"I believe so," Gabriel said and added mockingly, "though I'm not sure that you realise the full implications of what you're saying."

At Hillcrest, away from the anxious ministrations of his mother, Gabriel began to walk without crutches again, grim-faced at first, leaning heavily on Frances, from couch to verandah and back, to the border, down the lawn, further and further each day.

Sarah never knew where she would find them. She came through the gate into the flower garden one afternoon and saw them standing by the sundial, Gabriel's hand under Frances's chin, his head bent over her upturned face.

She retreated quickly into the vegetable garden and stood with her heart thumping strangely against her ribs, not seeing the feathery fronds of carrots, bushy rows of beans, or brightly coloured dahlias ready for cutting, but the picture of Frances and Gabriel by the sundial.

When at last she went back to the gate they were there still . . .

Lucy put her head round the study door. "I was about to throw out a dress," she said, "when I thought it might do for you, Sarah. It would have to be altered, of course, but we've got the dressmaker staying at the moment. Do you want to try it on?"

Sarah, struggling with a tricky piece of translation, thought for a moment that she had misheard. "May I, Gabriel? I'll come straight back."

As she went upstairs she heard him say, "You will take care, won't you? Don't give her anything too old," and Lucy reply, "You know you can trust me."

The dressmaker came twice a year to the Rectory, working, sleeping and eating in the little room at the back of the first floor. The sun never reached the window; even in summer one shivered going through the door. This morning the drugget was on the floor to catch the pins, the sewing-machine in position on the table. Miss Teape's breakfast tray, not yet cleared away, balanced precariously on the bed, remains of egg yolk crusty on the spoon. Hilda considered the dressmaker gave herself airs and resented waiting on her. Too good for the servants and not good enough for the family: that was Miss Teape's misfortune.

She smiled nervously as she held up the dress.

"Oh, Lucy . . ." Sarah stammered.

"I never really cared for it. It makes me look insipid, but you don't need to worry about that with your colouring. There doesn't seem much opportunity for wearing that sort of thing nowadays. It might as well hang in your wardrobe as mine. Put it on."

The white chiffon taffeta was lined with sarsenet and rustled magically as Lucy lifted it over Sarah's head. She had never worn anything so grown-up, a bodice cut so low.

"If we lift the skirt at the waist . . . " Lucy was saying, "and take in the side seams . . . "

Sarah looked down on Miss Teape's head as the dress-maker knelt on the floor measuring the hem, at the thin grey hair scraped back into a tight bun, criss-crossed with fierce black hairpins and wondered, as she wondered every time Miss Teape appeared, what it was like living in other people's houses, a week here, a few days there, sorting out wardrobes, mending, altering, making winter clothes in autumn, summer clothes in spring. Did Miss Teape have a house to go home to or were all her possessions contained here in this one small room? Perhaps you could make a story about her, or someone like her, a wandering Dutchwoman of a dress-maker ...

"I don't think it's too old," Lucy said. "Do you, Miss Teape?"

The dressmaker mumbled incomprehensibly through the pins that stuck out of her mouth like spider's teeth. When she stood up gusts of eucalyptus wafted round the room, an unpleasant reminder of chilblains and winter. Perhaps staying in north-facing rooms gave Miss Teape chilblains in summer, too.

Her hand lingered on the taffeta as she smoothed the skirt. How sad to spend your life making dresses of beautiful material if you could only afford unbecoming blue serge.

"Well?" Lucy said.

Sarah looked at herself in Mrs. Mackenzie's cheval-glass. "It's ... I don't know what to say."

"There's no need," Lucy said. "Looks are enough. Go and show yourself to Gabriel."

When Gabriel came downstairs in the morning he glanced first at Sarah's work and then sat down at the Rector's desk to read the newspaper. It was part of his routine; he was annoyed when it was disturbed.

"Where's the paper – have you seen it, Sarah? Ring the bell, there's a good girl."

Bertha stood in the doorway. Sarah knew at once that something was wrong. Bertha was nervous. Hilda was made of sterner stuff; considered her first duty to her mistress and would lie to anyone if necessary. Bertha could not lie. She avoided Gabriel's eyes as she said, "*The Times*, Mr. Gabriel? I always put it on the Rector's desk after breakfast."

"Then someone must have removed it. Find it, Bertha, please."

Bertha looked helplessly towards Sarah who was intrigued. Everything had its place in the Rectory; nothing was ever mislaid. "I'm sorry, sir, I can't."

"Can't? What do you mean, can't?"

Bertha hesitated. "Mrs. Mackenzie said you weren't to see it, sir. Not till she'd talked to you first. Only she had to go down to the village in a hurry to see to Nellie Gould's baby, so she said to put it away until she came back."

"Why should my mother . . . ?"

There was a long pause. "It was your regiment," Bertha whispered.

"Where's that paper?" Gabriel said.

Bertha shook her head. "Mrs. Mackenzie . . . in the morning-room." When Gabriel had limped from the room, she looked despairingly at Sarah and said, "Oh, Miss Sarah, whatever shall I do? Mrs. Mackenzie'll kill me, really she will . . ."

"Of course she won't, Bertha. He was bound to see it sooner or later."

"She wanted to prepare him, you see. It was the lists – all his friends, she said, in the latest push."

Gabriel was still in the morning-room when Sarah left the Rectory and at Hillcrest that afternoon they waited for him in vain.

The worst of the village casualties was Bill Roberts who was blinded in Sanctuary Wood and later had one hand amputated in a Chelsea hospital. Mary went up to London to stay in one of St. Dunstan's houses near the hospital while Jess came home to look after young Willy and the baby.

"You'd better go down and see what's what," Anne said to Sarah. "She always was an independent sort of girl, was Jess. Like Mary, come to think of it. It'll be easier for you to find out what's needed than me or your sisters."

But Jess appeared to have everything under control. The tiny kitchen was spotless, filled with the hunger-making smell of rabbit stew and Jess was changing the nappy of the baby squirming on the table with a dexterity that Sarah could only admire.

"We can manage," Jess said when Sarah tried to explain her reason for calling.

"Well," Sarah said, "if there is anything ..." She felt awkward, in the way, not sure whether to stay or depart.

Jess looked her in the face for the first time. "Still doing your lessons, are you?"

"I'm going to be a writer," Sarah said and at once wished the words unsaid. Why, when she had mentioned her hopes to no-one, did she have to blurt them out to Jess, unless it was in reply to some unspoken meaning or criticism underlying Jess's words?

"Oh yes? Well, I still want to be a teacher. You'd better get on with your work then, hadn't you? I've got plenty to do here."

Jess had changed. Impossible to recall the happy companionship of those lazy afternoons long ago before the war. "It's the war, of course," was how Mrs. Mackenzie excused the deteriorating conditions of life. It's all the war's fault, Sarah told herself, as she walked up the hill to Hillcrest, savagely decapitating dandelion heads against the trees of the hedge, but she knew that it was not.

The war was drawing closer, with air-raids more frequent on London and the south-east. Gabriel tried to persuade Frances not to go to London for the exhibition of her paintings but she was adamant.

"I must be there, you know that; at least for part of it. I have to talk to people."

"You hate having to talk to people you don't know. They'll all be philistines, anyway. They won't have any idea about painting or what you're trying to do."

"Some of them will. The critics ..."

"I'm more important than any critic. Please, Frances. Suppose a bomb fell on you. Or a Zeppelin."

She smiled. "It's a remote suppose."

"You're so ... it's all that matters, isn't it, that wretched exhibition?"

"Of course it isn't. You matter very much. But exhibitions don't happen every day and it is important for my work. You must realise that."

"I suppose I do. It takes a bit of getting used to, though – knowing that my only claim to fame will be the fact that I was a friend of that outstanding painter, Frances Purcell. Besides, I don't like having to worry."

"Now you know how I feel. Except that air-raids are nothing compared to the trenches. I shan't be running any risk."

"People get killed in air-raids. Nearly two hundred the other day."

"You're always telling me not to believe what you read in the papers."

He was irritable while Frances was away, unable to concentrate. He lay on the couch and sighed, picked up his book and put it down. He refused to walk, forgot what he had given Sarah to study, paid little attention to Gwen when she asked his advice about new shrub roses for the pergola.

"You need to be occupied," Gwen said and gave him the peas to shell for dinner.

The colander filled slowly. "I don't know why anyone bothers to cook peas. I never knew they were so delicious raw."

Sarah lifted her head from her book in amazement. "Haven't you ever shelled peas before?"

He was amused. "Of course I haven't. I don't suppose Lucy has either."

"She has over here sometimes," Sarah said.

The tea-tray shuddered in Annie's hands as she came out from the kitchen. "Mr. Gabriel! What *do* you think you're doing? A gentleman like you podding peas! Whatever would your mother say?"

"She'd be glad to know that I was making myself useful, I expect."

Annie put down the tray with a bang that made the china

jump, and snorted as she gathered the emptied pods into her apron. "I don't know what you're thinking of — and as for you, Miss Sarah, lying there with your nose in a book ..."

Sarah rolled over. "He likes doing it. Don't you, Gabriel?"

"I need the practice, Annie. When Frances is earning the money and I'm running the household — hasn't Frances told you that's what we're going to do?"

Annie sniffed. "I suppose you think that's funny. Marriage is no subject for jest let me tell you, Mr. Gabriel. I don't know, I'm sure — you can be as provoking as Miss Frances when you put your mind to it."

"Poor Annie," Gabriel said ruefully, as she swept into the house with as much dignity as an apronful of pea pods would allow. "I'm afraid I've offended her. There's no need for you to look so pop-eyed either, mouse."

"Oh, but Gabriel, are you — are you and Frances —?"

"Of course we're not."

Did he mean it? She wondered aloud to Annie: "They haven't quarrelled all summer, that I've heard, and Frances is ... well, different somehow."

"All sweetness and light. But we don't know why. Maybe it's her work. She always says she paints better when Mr. Gabriel's around. Leave off pondering, love, it's not for us to pry."

Frances returned early. Evensong was over and Mr. Mackenzie warming himself in the evening sun on the verandah before helping Gabriel back to the Rectory when footsteps sounded lightly on the front path and Frances came round the side of the house.

Gabriel's face lit up. "Frances! We didn't expect you until tomorrow."

"London was horrible. Full of nasty people and so dirty." She held his hand, her smile for him alone. "So I came home."

"How did it go? There was a very perceptive review in *The Times*, I thought."

"Yes, that was nice, wasn't it? It went all right — very well, really. I sold quite a lot. I still feel a fraud taking money for something I'd do anyway. However, I finished up feeling so rich that I went out and bought you all presents."

"Presents!" Sarah said. "Where? Can we open them now?"

"They're in my bag."

Sarah grabbed the bag, fumbled inside. "Which is mine? Goodness, this one's heavy."

Frances's expression changed. "Leave that. It's for Gabriel. I'll give it to him later."

"Oh no, you won't," Gabriel said. "I'm not waiting. Chuck it over, Sarah."

"No," Frances said quickly but it was too late. Gabriel had already torn off the wrapping paper.

"What is it?" Sarah asked.

He was suddenly sober. "Wire-cutters."

"Are they all right?" Frances asked, her eyes anxious.

He balanced them on his hand. "They're very good. Where did you get them?"

"The Army and Navy. I had a long talk with the man there. He said they were the best and he seemed to know what he was talking about. They're pretty powerful, he said – double lever."

Gabriel smiled. "I wish I could have seen you earnestly discussing the merits of different wire-cutters."

"He said the ones the army issue can't cope with German wire, they only cut British. That's not true, is it?"

"Yes, it is."

"But that means . . ."

"I keep telling you we might as well be fighting in the Crimea. Except that in those days we'd have been mounted."

Frances swallowed. "But that's awful, Gabriel. Some men couldn't afford . . ."

"I know."

She said after a moment, "He told me to tell you to use steady pressure. You'll break them if you try wrenching the wire apart."

"I'll remember that."

"I bought you two pairs in case you didn't."

He said mockingly, "What remarkably little faith you show in my competence."

She tried to smile. "I thought you might lose a pair in the mud or something."

His mouth curved up as he leant back and shut his eyes. The wire-cutters lay on the rug that covered his legs, their weight forming a hollow in the wool from which furrows and folds radiated and ran like miniature pathways across the green and brown checks.

Conversation died. They sat silent, all thought of presents waiting to be unwrapped forgotten, wiped out by the reminder of the world beyond the gate.

Slowly, purposefully, the shadow of the old pear tree crept across the lawn, taking with it the light and warmth that still remained from the summer afternoon, leaving behind a chill foretaste of the autumn yet to come.

Chapter Twenty

Once upon a time, in the years before Sarah remembered, the Mackenzies used to spend several weeks every summer by the sea at Budleigh Salterton. It was decided now that sea air would not only buck Gabriel up but help Mrs. Mackenzie, too. The strains of war life in England and the effort of nursing Gabriel had taken their toll: she was frequently grey-skinned with fatigue.

"Can't we go, too?" Sarah asked Frances. "The Mackenzies' landlady must know somewhere we could stay. Please, Frances."

"What makes you think the Mackenzies would want us with them?"

"You know Gabriel would like you there."

Frances sighed. "I'm sorry, Sarah, really I am. I'd love to go but we can't afford it."

She remembered the feel of sand under her feet, the smell of salt, the cry of the gulls. "Just for a day or two?"

"There's still the expense of the journey."

Life without Gabriel was empty, dreary, the ten days stretching into the distant future like months or years, yet surprisingly they ended at last and Gabriel and Lucy came over for tea on the Hillcrest lawn as if the routine had never been interrupted.

"It made a nice break," Lucy said. "We were afraid it would be very quiet but it wasn't at all. You wouldn't have known there was a war on, in fact. It was quite extraordinary."

"Full of high society who couldn't get down to Biarritz, on account of the nasty fighting." Gabriel made a face at Sarah. "The poor things."

"It hadn't changed from when we were last there. Bathing machines on the front, old Gooding fussing over the diving raft. Comforting, somehow, when everything else is upside down."

"You'll never guess the latest craze," Gabriel said. "Collecting moss from Woodbury Common. It gives purpose to your picnic, you see. You may be having fun but goodness me, you're being patriotic, too. There must be some clever epigram about rolling stones and ladies gathering moss, but I can't think of one at the moment. Can you, Sarah?"

Sarah thought for a moment. "No. What's the point of collecting moss anyway?"

Lucy gave Gabriel a disapproving glance. "It's for wound dressings in France, I think. I expect Julia would know."

Sarah never doubted that the holiday had been a success until the evening Lucy appeared at Hillcrest's front door.

"Do you mind if I join you? I'm not interrupting anything? I can't stand the Rectory a moment longer – Father and Gabriel bickering all the time."

"Your father and – you can't mean it?"

"I don't know what's got into Gabriel these days," Lucy said. "He's been so bad-tempered, ever since we went away. They're arguing about Ireland now, on and on, and getting so angry. As if it'll make any difference."

"I've never understood why Gabriel gets so hot under the collar about Ireland," Gwen said.

"He's always supported Home Rule, you know, even before the war. It was the one thing he and Father never agreed on. And then that friend of his — you know him, Frances, the one who was killed at Arras . . ."

"James."

"That's right. James was always very wrought up about the Irish question; I think he was Irish as a matter of fact and I know he was in Dublin during that business last year. It must have rubbed off on Gabriel."

"Can't your mother . . .?"

"She makes it worse. You know Mother — if it'd been up to her she'd have sent in the gunboats long ago. Gabriel really does see red when she gets going."

"It's not Ireland," Frances said, "but the mood he's in. I wish you'd never gone to Budleigh Salterton. He was all right until then. Now all he can think about is getting back to France."

"He's told you that, has he?" Lucy said miserably. She glanced at Sarah who was pretending to read a book. "Isn't it about time she went to bed?" And after Sarah had left the room, protesting bitterly but careful to leave the door ajar, Lucy said, "I didn't want to talk about it in front of her but it's true. He's determined to get back to the front as soon as possible. Within the next three weeks he says."

"But that's ridiculous!" said Gwen. "He still gets quite a lot of pain, doesn't he? He can't walk properly yet. I thought the muscle . . ."

"Why do you think he's started on these expeditions?" Frances said. "He's worked out a training programme for himself, pushing himself to the limit. Beyond the limit, in fact. I thought I was going to have to carry him from Dunkery yesterday."

"He can't really want to go back, can he, Frances?"

Sarah crouched on the stairs, head pressed against the banisters, angry with the grandfather clock whose measured tocks were louder than the voices below.

"He told me once that just the thought of the trenches made him feel physically sick. That was after the Somme. It must be worse now. He says he can't stand England, the way people complain, hate the Germans. He wants to get back to his regiment. It's only since Budleigh Salterton; he was happy enough before that. Whatever happened, Lucy?"

"It was awful, really it was. The place was full of fair-haired little boys for one thing; we kept on seeing Antony everywhere. He was always an absolute horror on holiday, you know, falling off rocks, nearly drowning himself in pools. You can't help remembering that sort of thing. Then there were the terrible women Mother played bridge with – going on about their husbands and sons. You weren't accepted unless at least one of the family'd been killed or wounded – and the more mutilated the better, it seemed to me. They were all so cheerful about it, too. It was unbelievable really. You can imagine Gabriel's reaction. He was dreadfully rude to some of them and that upset Mother. He was ... I don't know. He's said some horrid things to me, as a matter of fact. What a useless life I lead, why don't I follow Julia's example – that sort of thing."

"How does he think your parents would manage without you?"

"He says Mother's migraines are just nerves."

"You mustn't let him upset you," Frances said. "He has these moods. He doesn't mean half he says."

"I kept on thinking: if only you'd been there, too. He's all right when you're around, Frances. Gwen and I could have laughed at the dreadful women. And Sarah would have cheered us all up ..."

"We did wonder. It was the expense – the way prices go up ... and we have to think of the future with Sarah."

"If that was all, you know Mother and Father would have gladly paid."

"Don't be absurd, Lucy."

"I do wish you weren't so prickly about money, Frances. They'd *like* to be able to help sometimes but you never give them a chance."

"Well, if I'd thought ... Sarah could have gone. She was desperate to go."

Annie passed by the foot of the stairs and caught sight of Sarah crouching in the gloom. "Bed," she mouthed and pointed upwards.

Sarah looked out over the darkening churchyard and thought, I could have been part of a family. Living together, eating together, walking, talking, all day spent together. Mother, father, brother, sister. Gabriel and I. No-one would have known we weren't a proper family.

Gabriel never mentioned the war to Sarah. They talked about her future; never his. He kept his deeper feelings hidden from her except for a brief glimpse one afternoon.

He was helping her net the ripening figs against the depredations of birds and wasps, supporting the ladder and passing up muslin squares one by one as she perched precariously on the topmost rung, when quite suddenly he said, "The garden of Eden must have been like this, don't you think?"

The ladder rested against the tree, shifting slightly whenever she moved. It took all her concentration to tie the muslin bags firmly round the fruit without overbalancing.

"I hadn't thought about it."

"They wouldn't have known what it was like outside, either."

She glanced down at him. He stood gazing with eyes half-shut into the distance, past the hens scrabbling on the path to the stables, past the walnut tree and copper beech beyond, past the magnolia ...

"If only they'd known ... "

She went on working, covering each fig that she could reach, making flowers of the fruit, white muslin flowers against the dark glossy leaves. She felt inadequate, not sure what he meant or how to reply, sensing that he needed comfort of some kind but in her inexperience unable to give it.

When at last she climbed slowly down he gave her a faint, shamefaced smile but said nothing.

They walked silently back to the verandah, the grass warm and soft under the soles of her feet. The garden drowsed in the summer sun. Cabbage whites fluttered over the nettles that grew thickly between wall and greenhouse; bees hung round the herb garden and over the sedums. In front of the pale spikes of the delphiniums' second flowering, blue-black poppy heads flopped onto the red path.

It's my garden of Eden anyway, Sarah thought. Please don't ever send me away.

The day before he left Gabriel presented Sarah with a long, thin parcel. "For you, mouse."

"For me? But Gabriel, you've given me so much – Quiller-Couch, Roget . . ."

"Yes, well. This is different."

She unwrapped it with trembling fingers and gazed incredulously at the spray of flowers and leaves, delicately outlined in silver, filled in with clustered pearls. "Oh, Gabriel!"

He looked down on it, his face tight. "It's not a very suitable present for a child. You'd better ask Frances to look after it until you're older. I thought . . . well, you could have it as an early present for your twenty-first."

"That's not for years and years yet," she said laughing, holding the necklace against her, the silver cold and thin against her skin. "It's beautiful."

"I don't know when you'll be able to wear it," he said doubtfully.

"With my new dress."

"When are you going to wear that? I can see that I shall have to take you out to dinner myself one of these days."

His imminent departure made her bold. "You promised to take me to the opera once."

He smiled. "Did I really? The things I let myself in for."

But the opera was far away in London; it was Frances he

took out to dinner in Sir James's borrowed car. Of course it would be Frances. Why should he take a child, a mouse? Except that she no longer felt like a child. She would have to smile at them both, look forward to next year, wait until she was grown-up, the war over...

What was it like dining in a restaurant? What did you talk about during those long hours with a man, alone? How would it feel, to have Gabriel's arm round her, holding her close, dancing ... while the notes of the trumpet swooped and soared.

But Bill Roberts would never play *Ramona* again and by this time tomorrow Gabriel would be on his way back to the front.

Chapter Twenty One

Hopes that Geoffrey might come home on leave before Gabriel's departure were not fulfilled. A shortage of officers, renewal of activity round Ypres – there was always some reason why leave was deferred. It was October before Geoffrey was once more back in England.

"I don't know how we are going to entertain him," Mrs. Mackenzie said. "October! It's too late for tennis, and tea parties are so difficult nowadays. He'll find it so dull. That must be why he is staying with his friend – Patterson is it? – in Surrey. One would have expected him to come home first."

"No doubt they hope to take in a show or two," Mr. Mackenzie said mildly. "I believe the Pattersons' place is within comparatively easy reach of London. You must allow him a little pleasure, my dear. As for entertainment, I'm sure he will be happy to be home and able to sleep in a proper bed again."

"Do you think we shall find him much changed? I did think, didn't you, that Fergus Donne had become very cynical the last time he was home on leave."

Apart from the addition of a moustache, Geoffrey looked much the same when they met him at the station, though thinner perhaps and tired. The girls, if not Mr. and Mrs.

Mackenzie, assumed that the air of fatigue was due to London night life and hoped to hear details when they got him on his own. They were disappointed: Geoffrey was unwilling to tell them anything. It was soon apparent that he had changed considerably. He no longer cared what people thought, had become fidgety, irritable, even rude.

"You must let me see your work," he said to Frances when he came over to Hillcrest. "I think you're marvellous, really I do. The whole world goes up in flames but never mind, Frances Purcell paints on. Bit like Nero, isn't it?"

Mrs. Mackenzie was more than once reduced to tears.

"It's too bad of you, Geoffrey. Why do you have to behave so badly when we looked forward to your coming home so much? We're doing our best to give you a good time, but you must realise that entertaining isn't easy, with all the shortages. The newspapers tell us we shouldn't serve tea at all these days."

"You know I've always loathed tea parties and I never could stand Jane Lynch. Or any of those females you had yesterday."

"You used to like Jane. She's a nice girl, of good family. There was no need to be so — so outspoken."

"Girls like her make me ill. They twitter on about their war effort and what are they doing when you come down to it? Flirting with convalescent officers, that's all."

"Jane's very dedicated. She puts in long hours at the hospital. It would have been kind to show some interest."

"Long hours — rolling bandages, dishing out tea! My God, Mother, you're living in a bloody fairy-tale here. It's not real. If Jane Lynch wants to do war work why the hell doesn't she do what Julia did and go out to France?"

"That's quite enough, Geoffrey. I don't know what's come over you. You don't seem to consider my feelings — anyone's feelings. Sister Vicary complained to me yesterday about your rudeness. How *could* you ask her if she'd served in the Crimea? And another thing: I don't want to criticise when you're home for so short a time but your language has become very ... well, coarse. In front of Sarah, too. And talking of Sarah, I don't know why you should spend all your time with her. She's only a child."

"He's helping me," Sarah said quickly, feeling hot and uncomfortable. She wanted to protect Geoffrey but felt sorry for Mrs. Mackenzie at the same time. If only they wouldn't quarrel in front of her.

"You surprise me, Mother. It's part of Sarah's war effort; don't you think I should encourage her? Rather more worthwhile than making eyes at wounded officers, surely? There must be precious little ammunition left on the western front after the last few months, I'd have thought. I can't imagine how horse chestnuts can do much damage to the Hun but if there's a chance of it I'd rather help Sarah than listen to those damned girls going on and on about their conquests. Come on, Sarah, let's go."

He came over to dine at Hillcrest. The evening was not a success, though that was not altogether his fault. It was the first time Dora had had to serve anyone other than the Purcells and the occasion so unnerved her that she dropped a plate of vegetables onto the floor. "If anyone can make anything of that girl, it'll be Annie and Hillcrest," Mrs. Mackenzie had said when sorting out the school leavers, but it was hard work and the Purcells often thought longingly of Florrie, the best girl they had ever trained, who had gone on to Sir James and Lady Donne's London house where she not only earned more than the Purcells could ever hope to pay but had prospects as well.

Geoffrey looked down at the tomatoes splattered over the carpet, turned deathly white and became so uncommunicative that the Purcells, anxious for news of Julia who by now had been in France for over a year, nearly screamed with frustration.

"For goodness' sake, Geoffrey! Can't you tell us anything? It's so difficult finding out from letters."

"I'm sorry. She's just the same. Well, no, I suppose she isn't. We've all changed – not you, I suppose, but we have."

"What do you mean?"

He tipped his wineglass so that the red liquid slurped to and fro. "She looks the same, anyway."

"Is she happy, do you think?"

He stared. "What a ridiculous question! It's pretty grim, you know. Jane Lynch should try it. Or you lot. You wouldn't last a day, Frances. Keel over on your back after the first five minutes."

Frances was surprisingly unruffled by Geoffrey's rude tones; she sat watching him with an anxious, oddly maternal expression on her face.

"I didn't mean that, exactly. I meant, is she satisfied with what she's doing? She doesn't want to come home? Does she have time to paint, do you know?"

"For God's sake, Frances! Get out the easel between doing the dressings, do you mean? Half the time she doesn't get enough sleep to keep her going."

Frances flushed. "I'd have thought painting might help her relax, take her mind off things."

He twisted the stem of the wine glass between finger and thumb. "Yes, well. She did draw quite a bit in Normandy, as a matter of fact."

"Tell us about that," Gwen said. "Why didn't Julia say you'd been on holiday together? She told us she met you in Paris; why did she never mention Normandy?"

He shrugged.

"Where did you go?"

"All over."

"What's Normandy like?"

He considered. "I like the Quantocks better."

"What did you do?"

"Walked."

"And?"

"Just walked."

He was nervy as well as taciturn. His fingers played continually with the buttons of his jacket and there was a twitch in his cheek. He jumped when Gwen stood up to put a log on the fire, knocked over his glass and tipped elderberry wine in a bright spreading stain over the tablecloth.

"It's his nerves," Frances said after he had departed, having spent the greater part of the evening glowering into the fire. "I don't think he's safe to go back. He's as likely to shoot one of ours as one of theirs, if you ask me."

"He's all right during the day," Sarah said defensively. "He talks more than he did tonight. He's fun to be with."

It was true. She enjoyed the afternoons she spent with Geoffrey, tramping round the countryside in search of chestnuts. They had soon exhausted the local trees – not a fallen leaf remained unturned, not a chestnut left on the ground. In the distant lanes between the villages they found trees unvisited, with a hidden harvest of conkers buried in the layers of papery leaves on the ground.

"I made you some chestnut furniture once," Geoffrey said. "With pins for chair legs. Do you remember?"

"It's still in the back of my cupboard, I expect," Sarah said. Untouched, she thought guiltily; she had never cared much for dolls or dolls' accoutrement.

"If we had that motor bike I'm going to get after the war, you could sit on the back and we'd go to every chestnut tree in Somerset. No-one else would have a chance then."

"What do you think it's like, riding on a motor bike?"

"Tremendous. Really it is. I borrowed one once, to see Julia. Fifty-five miles I went, on a borrowed bike. How about that? I was like a rag at the end of the day. It was worth it though. We went to Paris-Plage and ate French pastries and bathed in the sea."

"I thought Paris was inland."

"Paris-Plage, silly, not Paris. It's near Étaples. You can get a train, or a tram, but we walked through the woods. The sand's marvellous, miles of it. Much better than St. Audries."

They threw leaves at each other, challenged each other to stick-throwing contests. He seemed no older than she was. When they talked it was always of the past.

"Do you remember that walk to the sea ...?"

"Shouting Coleridge all the way ..."

"That was the first time I'd heard of the man from Porlock."

"And the cockerel outside the barn ..."

"We were going to go back every year."

"We still can ... when the war's over."

"Annie was furious when we got back."

"So were Mother and Father. You should have heard them going on at Gabriel."

"I don't see why they should have been cross."

"Oh, I don't know. We'd ruined your reputations, I expect. Don't you think you've got enough chestnuts now?"

"It's David Shattock. He thinks girls are silly. I can't let him win."

She wanted the prize that Sir James had offered for the weightiest collection of chestnuts – a copy of *The Red Cross Story Book*, full of stories by writers who were at the front – but more than the prize Sarah wanted to see the grin wiped off the face of that smug little Shattock boy. He had told her that girls were cissy: she would show him who was best.

It seemed a funny way to help the war, collecting conkers, and impossible to believe that the familiar, glossy playthings could be used to cause the sort of wound that had kept Gabriel at home for so long.

"If all the children in England collected all the chestnuts ..." the Rector had said from the pulpit. Chestnuts could be used instead of grain to make ammunition. With all that extra grain for bread who need worry about German submarines in the Atlantic?

Sarah surveyed the pounds of chestnuts, dull now and wrinkled, laid out in the parish room and thought that the war must surely end soon and Gabriel come home once more with Julia and Geoffrey.

Chapter Twenty Two

Sarah's thoughts were on her lessons as she hurried home through the churchyard; she almost missed the figure by the stone tomb. Lessons were not such fun these days; Mr. Mackenzie sad and distrait since Gabriel's return to France. If only Gabriel might be wounded again, Sarah was thinking – a little wound but enough to bring him back to the Rectory – when out of the corner of her eye she caught sight of movement. She turned.

"Geoffrey! What are you doing? Are you all right?"

He wiped his mouth with the back of his hand. His face was pale green, shining with sweat. "N-no, not really. I – I've just been sick."

"You look awful. You're not going to faint, are you? Here, come and sit down for a minute."

She pulled him over to the porch, but he refused to sit on the stone bench, muttering something unintelligible about people passing. Somewhat apprehensively, she dragged him into the church itself.

"What's the matter? Is it something you've eaten?"

He shook his head. "I couldn't face lunch."

She sat beside him, uncertain what to do. Unless she hurried she would be late for lunch herself, but she could hardly leave him here on his own, ill.

"What's the matter?" she said again. "Are you sickening for something?"

"Oh, if only I were. Nothing's the matter. Except I can't get the noises out of my head. It's just ..." Tears filled his eyes and rolled down his cheeks. "Oh, Sarah, it's so awful ... you can't imagine how awful ..."

She had never seen a man cry before. She was shocked as well as frightened. Her stomach knotted. "What's awful, Geoffrey?"

It was the most terrible thing that had ever happened to her. She sat in the familiar church, as she had sat every Sunday for as long as she could remember, looking at the solid stone-work, the delicate tracery of the ash-grey rood-screen, and listened to details of such horror that could only have been conceived in some mediaeval imaginings of hell. I won't listen, she wanted to say, don't tell me, I don't want to know. But she sat silent, while he went on and on, his voice expressionless, his face blank, while the tears ran down his cheeks and dripped onto his jacket.

"In England it's ... it's ..." He stopped, started again. "Once, when I was at Patterson's, we all went down to the coast. For a picnic they said, but really it was so that the girls could hear the guns. They stood on the cliff and chattered. Just think, they said – they're banging away for England. Isn't it splendid, they said. You could hear them quite clearly. The guns, I mean. I hear them all the time. Even in my sleep. They go on and on."

He looked as if he had been unable to sleep for weeks. The greenish pallor had left his face but the tan sat oddly on his skin, like an outer layer put on for protection. She held his hand tightly in hers. It lay cold and motionless, like a dead man's.

"Sometimes I think I'm still there. It isn't just the noise. I see bodies all the time, too. There were bodies all the way down from London, lined up along the railway line. Dead men. Their eyes had gone and their lips were pulled back, laughing at me. They were quite dead, I knew that."

He bore no resemblance to the boy with whom she had searched for chestnuts. He looked at her as if he had never seen her before and said conversationally, "I remember such odd things. Half-eaten bread ... we took a Hun trench once – oh, ages ago now. They were in the middle of a meal and left in a hurry ... there was some bread left on the table. I keep on seeing it. Rough brown stuff. We had to kill one of them, you know. It was him or us, so Patterson shot him. He died there in the trench. The Hun, I mean. He kept asking for his mother. I was the only one who understood so I had to pretend – He wasn't very old, about the same age as Antony ... It's all so pointless, so stupid. What are we fighting for anyway?"

"I don't know. God, I think."

"There can't be a God – He wouldn't let such things happen. Or perhaps there is and He's helping them. They think God's on their side, you know."

"England then. We're fighting for England."

"But England's not – it's not what I thought it was."

She could think of nothing to say or do that would help. In the hollow emptiness of the church she prayed wordlessly for some means of comfort. None came.

He stirred at last and said, sounding very tired but more normal, "I'd better go. I'm sorry, Sarah, I didn't mean ... Forget it, will you? Don't tell Julia."

She remained in the church for a long time. It was part of her world, the world of church, Rectory, village and Hillcrest; familiar, well-loved. But it was self-contained, too, and isolated. In Huish Priory even the guns were too far away to be heard.

It was not possible to talk to Frances, when Gabriel had so recently returned to the front; nor Annie, who had one brother at Plugstreet and another fighting the Turks, supposedly so much worse than the Germans ...

Gwen was out in the vegetable garden, muffled up against an autumn afternoon that had become cold and grey. She looked up when Sarah approached, smiled, and continued dibbling holes at regular intervals in the earth of the cold frame. Sarah sat on the brickwork and watched her separate the seedlings and gently drop each one into its own place.

"I don't think Geoffrey wants to go back," she said at last.

"Neither did Gabriel. Oh, I know he insisted on going back before he was properly fit but he wasn't very bright when the day came, was he?" She glanced at Sarah. "Worse than that?"

"Well ..." She searched for words but in vain. She knew then that she could never tell anyone, would never be able to repeat the terrible things that Geoffrey had told her. "Yes."

Gwen pushed her fingers into the soil and made it firm round the plants. Pieces of red earth stuck to her knuckles and nails. The Mackenzies had beautiful hands, smooth-skinned, well cared for. Purcell hands were always stained, with paint or ink or mud.

"Can't he talk to his parents? You always think that must be one of the advantages of having parents. Surely Mr. Mackenzie ..."

"Geoffrey says God doesn't exist."

"Oh, lord. That's not much good then. And Mrs. Mackenzie would go on about doing one's duty and sacrifice and those who've gone before helping from Up There. What about the Germans Up There, I always wonder. Are they helping too? Their own side, presumably. In which case it gets more and more like Antony's precious gods every day, doesn't it?" She sighed, gestured to Sarah to get off the cold frame and pulled down the top. "I'm sorry, Sarah. I know it's awful but he's got to go back, hasn't he?"

Of course he had. Just as Sarah had to go over with Gwen and Frances to the Rectory that evening for a farewell dinner of Geoffrey's favourite food, very little of which Geoffrey himself managed to eat. His eyes across the dinner-table reminded Sarah of a petrified rabbit caught on the harvest field.

"I thought you and Mrs. Mackenzie were getting on better these days," Gwen said to Frances as the three sisters walked soberly home. "What on earth possessed you this evening? Why do you have to behave so badly?"

"She made me wild. On and on – Antony's supreme sacrifice, Gabriel's M.C. What does she expect Geoffrey to do – go out and get the Victoria Cross?"

"She doesn't mean to do it. She's frightened."

"Don't you think I'm frightened? I don't go on like that – on and on about how brave Gabriel is. Come on, Gwen, you know Gabriel only got the M.C. because he lost his temper. He was so angry he didn't think what he was doing. If it hadn't come off he'd probably have been court-martialled."

"But what did you hope to do by contradicting her all the time?"

"I don't know," Frances said wearily. "Stop her talking about the other two, I suppose. Talk about Geoffrey himself, for a change."

"Poor Geoffrey. He looked awful, didn't he?"

Sarah's thoughts were far away. Was Gabriel staggering heavy-laden up the line at this moment, balancing precariously on slippery duckboards, in danger of being knocked off, blown off, drowned in mud? If he looked upwards would he see the stars that she could see, twinkling eternally in the darkness? Please God, look after him. I'll do anything You want for ever and ever so long as You keep him safe.

Afterwards she thought that she had known that evening at the Rectory that Geoffrey would never come back. But Mrs. Mackenzie, with the telegram in her hands, refused to accept the fact of his death. Geoffrey had been posted missing, unlike Gabriel who had been believed killed; he must therefore be somewhere among the wounded or on his way to Germany as a prisoner-of-war.

"It's a matter of waiting," she said, and continued to believe it, long after the arrival of the letter from the company major admitting that there was no hope of Geoffrey's survival.

Every night Sarah lay in her soft, comfortable bed and watched Geoffrey drown in the bloodstained water of a shell hole, saw him sucked down into the death-smelling mud; helpless herself saw him lying helpless between the front lines, slowly dying. How many days did it take, how many nights?

Geoffrey's death was worse than Antony's for it brought her face to face with the reality of Antony's loss. Before Gallipoli and ever since, Antony had been sailing forever in her imagination on mythical seas: Jason among the Argonauts, Agamemnon with his Greek princes, Odysseus exploring the far islands. Never once had she thought about the scene on the beaches. Now she thought of little else.

Had she really believed that some time in the future he would return to her, older, weather-beaten, experienced? Had her work, her absorption in books, been like Penelope's tapestry — a refusal to face the truth?

She reached a grey place in her mind. She worked, went about as usual, behaved as she had always behaved, but inside she felt separated from all that she knew and loved. It was as if she were walled round by a glass barrier, invisible but surely there. She walked under the trees whose chestnuts she had collected with Geoffrey, through the churchyard that Geoffrey and Antony had known, looked out on views that they had loved. The curves in the road, the spinney, stream and distant Quantocks had all been here for years past and would remain for years to come, long after Geoffrey and Antony were nothing more than forgotten names on the parish memorial, long after she herself had joined her own mother in the churchyard.

For the first time she became aware of her own mortality, terrified by the possibility of her own end. How was she to know whether an ache or sudden pain might not be the first sign of some fatal illness? When she went to bed at night how could she be certain of waking in the morning?

Some evenings she hesitated by the grandfather clock on her way to bed, hands wet, heart thumping at the thought of the waiting nightmares, the possibility of everlasting darkness, legs shaking too much to carry her up the remaining stairs. For there were not only the sleeping nightmares to be endured, watching Geoffrey die over and over again, but the waking ones, too, when she had to face the knowledge that she was glad Geoffrey was dead, glad that it was Geoffrey who had been killed rather than Gabriel.

Such distress could not be concealed for ever. Her sisters became concerned, Annie anxious.

"It's nothing," Sarah said. "Bad dreams, that's all. I had them a lot when I was little, if you remember."

"That was your mother's fault," Annie said. "Letting you read whatever you wanted, fairy-tales and suchlike. I told her 'twas a mistake. Grimm by name, grim by nature, that's what I always said."

"Are you working too hard?" Gwen said.

"Mr. Mackenzie doesn't think so," Frances said. "I don't think it's just the nightmares, is it, mouse? What's the matter; can't you tell us?"

Tears came into Sarah's eyes at the pet name. "I don't know. I can't explain."

"Growing pains, that's what it is," Annie said. "I remember Miss Frances at her age. Tears and tantrums all the time."

"That was different. Mother and I were arguing about art school. It had nothing to do with growing up. You do want to go to university, don't you, Sarah? You mustn't let us force you into it."

1920. That was when Mr. Mackenzie said she would go to Oxford. Impossible to believe that she would still be alive in 1920.

"I want to go."

"Art school or growing pains, 'tis all the same," Annie said.

"We had problems with Miss Julia, too. 'Twas Miss Gwen was the easy one."

"She dug the vegetable garden instead," Frances said. "I remember one winter she turned it over half a dozen times."

Gwen smiled. "Which is why it's so fertile now."

"Ah well," Annie said, "I daresay things have to work through the system. How about Miss Sarah doing a spot of digging? Reading and writing all the time – strikes me she could do with a bit of going about."

"Why not?" Frances said. "Not digging though. We'll walk, Sarah."

It became routine. Whatever the weather the two of them set off after lunch every day along the narrow lanes round the village. As the days lengthened they walked further afield until the lighter evenings allowed them to reach and explore the Quantocks. Frances was an easy companion. There was no need to make conversation. Some days they walked for the whole afternoon in comfortable silence, on others they talked all the time they were out.

The ten years' difference in age and Frances's talent had always been a barrier between them. For almost the first time Sarah began to see Frances as someone not much different from herself. It had never occurred to her, for instance, that Frances might be lonely, that she might miss Julia's warmth, her easy-going and cheerful nature, as much if not more than the rest of them, though she had realised that the gap left by Gabriel's departure must be very much worse for Frances than for anyone else.

As the dreary winter of 1918 slowly gave way to spring they went to the Quantocks every afternoon, following twisting sheep tracks, trampling down the crackling stalks of dead bracken, looking out over the vale spread out below in a patchwork of ploughed earth, bare orchards, blue-green fields of winter cabbage.

Every corner of the Quantocks held memories of Gabriel for Frances. Each thicket, every patch of gorse reminded her of some particular time: a pause to talk or draw, watch a blackcock or listen to a lark, to have a picnic. She found a nest, first discovered by Gabriel years ago, being refurbished once again for this year's family. She talked to Sarah, reminisced gently and with love. It was difficult to see in this Frances the girl who so provoked Gabriel's mother.

To Sarah, walking in the hills where he was happiest, Gabriel seemed very close. Sometimes she was sure that if she could only turn her head quickly enough she must glimpse the young Achilles she still remembered, striding along the trackway in the distance.

Long hours of exercise in fresh air brought deeper sleep. The nightmares became less frequent, the fears less stark. Despair faded but so gradually that she was unaware of its disappearance.

One morning, pegging tea towels out on the line for Annie, she realised that she was free. She looked round the familiar yard, at the apricot-painted outbuildings with their green slate roofs, at the glistening cobbles and the robin watching her with head aslant from the water butt and thought, I'm not unhappy any more.

It was no more positive than that, yet strangely she was comforted, for now she realised that even the darkest shadows pass in time.

1919

Chapter Twenty Three

The Blackdown Hills had disappeared behind the rain, the outline of the boundary trees become smudged and grey. Soon even the rose garden would have vanished into mist.

Julia glowered at the shadowy garden. "Why today of all days?"

"We do need the rain," Gwen said, not lifting her eyes from her painting. "Come away from the window. You won't stop the rain by looking at it."

"I don't know why you're worrying," Sarah said. "Frances knows the Quantocks like the back of her hand and Gabriel knows them even better. They'll have found shelter somewhere."

"It's not fair," Julia said. "His first day home after so long and it has to be a rotten one like this."

"Life isn't fair," Gwen said. "You should know that. They'll be all right. As long as they're together –" She glanced at Julia. "What do you think'll happen?"

Julia sat down at the table. "I don't know. You can't tell with Frances, can you?"

Frances thought the war had changed Julia; she said that Julia had become hard and uncaring during the time that she had been away. Looking at her sister now, Sarah couldn't agree. She thought Julia looked sad and vulnerable; exhausted, as though she no longer possessed the strength that life required in order to endure. It'll get better, Sarah wanted to tell her, I know it will. But perhaps, when the man you loved was somewhere in Flanders with not even a grave to lay flowers on, when you could no longer paint because of the nightmares that appeared unbidden from your brush, perhaps it never did get better.

Gwen frowned down at the dahlia petal she was colouring. "I wish you'd try again, Julia. Why don't you start with one leaf, say, or a flower? You must get back to painting; it's such a waste if you don't. And you know what Dr. Milne said, that it was just the effect of the influenza and being tired out after the war and in time . . ."

"I'll try sometime. Not today. Next week, perhaps."

There was a shout from the kitchen.

"They're back."

In the scullery Gabriel and Frances had dropped their rucksacks on the floor. They stood shaking the rain off their clothes. Water formed into puddles, the puddles merged into a small lake on the flagstones round them. Their faces were shining with wetness, drops of water suspended on their lashes. Beyond the open door the rain bounced up from the cobbles.

"Lord love us, Mr. Gabriel," Annie exclaimed, "what's she been doing to you?"

"Don't be silly, Annie," Gabriel said shortly. "It's raining. We got wet, that's all."

"You're more than wet," Annie said. "Shrammed, you look. Let's have your jacket now."

Frances shook out her coat. "We thought he'd better dry off here. You know Mrs. Mackenzie – she'd have a fit if he went home like that. I'll go and change; I won't be a moment. Get him a towel, would you, Annie?"

"Into the kitchen with you," Annie said. "Get yourself in front of the fire before you catch your death. Boots off, please,

Mr. Gabriel. I'm not having Quantocks mud on my clean floor. Leave them here. Lily will see to them."

"You're worse than my mother," Gabriel protested, but he removed his boots and meekly padded into the kitchen on stockinged feet.

"Let's have your shirt," Julia said. "How did you get so wet? Couldn't you find any shelter?"

"It didn't seem so bad up there; it was the wind driving at us on the way home that did it. Don't look at me like that, Julia; you're not in hospital now. I suppose you want my trousers too?"

She laughed. "In a houseful of women I think we'll leave you with those. Sit close to the fire. I'll get something to put round your shoulders."

For a moment, watching Gabriel as he sat forward in Annie's chair, staring into the fire, Sarah thought he looked . . . defeated. Then he turned his head and saw her and she realised that the thought had been a trick of her imagination.

"Well," he smiled, "two years make quite a difference. Grown-up at last."

He was very lean; thin almost. She could count his ribs. His skin was honey-coloured, his face darker above the line of his collar. He shivered violently.

"Would you like to make me a hot drink? I don't think I shall ever be warm again."

She peered into the stock pot. "There's some soup. Or would you rather have tea?"

"Soup would be splendid."

She ladled stock into a pan, added a handful of vegetables left over from lunch and heated it up.

"I don't know why you had to go so far," Gwen said. "There were plenty of storm clouds about this morning. You could have had a picnic up in the stables – spent the day together. No-one would have known – we wouldn't have said."

"You know how I am about the Quantocks. I never feel at home until I've been up there."

Frances came into the room. Her hair hung thickly down her back, made two-toned by the rain, dark and light, wet and dry, the soaked ends frizzing up into tiny curls. Her blouse was already damp where wet hair had touched it.

"Here's a towel. Let's dry your hair." She stood behind Gabriel, her face absorbed, the towel white against his skin.

"Aren't men lucky?" she said. "Short hair takes no time at all. I suppose you don't need to bother really."

He reached up and caught her hands in his, leaning back against her with eyes shut.

"Oh, Frances."

She leant over, her arms round his neck, and bent her head over his so that her cheek rested on his hair. Her expression was tender, surprisingly sad.

Sarah stood with the mug of hot soup in her hand and stared at her sister: Frances, so sparing with her caresses, who rarely touched in public and never embraced.

"How much longer before you get out of the army?" Gwen asked. "Will you have time at home before Cambridge?"

Gabriel opened his eyes. "I'm not going back to Cambridge."

"Not –? Where are you going?"

"Ireland, probably."

"Ireland? Ireland? Whatever for?"

"The army is ... rather preoccupied with Ireland at the moment," Gabriel said, almost apologetically.

"The army? You don't mean ...?"

"Yes."

"You *can't* mean ...?"

"I've decided to stay on."

They stared at him incredulously. Gwen said, "He's joking, isn't he, Frances?"

She came away from his chair. "Apparently not. I think he's mad."

"What did your parents say?"

"I haven't told them yet."

"Oh," said Julia. "Oh, I see . . ."

Frances turned on her. "Be quiet!"

"I'll have to break it gradually. They may be upset."

"Upset!" Gwen repeated. "That's the biggest under-statement I've ever heard. Haven't you given them any warning?"

He flushed. "No."

"Of course he hasn't," Julia said, "and you know why . . ."

"What made you put vegetables in the soup, Sarah?" Frances said sharply. "You can't drink carrots out of a mug."

"I thought it was a good idea," Gabriel said. "Nice and filling. I should have asked Annie to see to our picnic. Cook has no idea what walkers need in the way of sustenance. Talking of Annie, how about some fruit-cake? Don't tell me you haven't got some hidden away."

Julia fetched the tin from the larder. "Gabriel, you're not serious, are you? Ireland, of all places."

"Don't let's talk about it," Frances said. "Gabriel's gone on and on all afternoon. If I hear another word I shall scream."

There was silence.

"How was the army of occupation?" Julia said at last.

"Interesting."

"And Cologne?"

"Horrible."

"We heard tales of dances and all sorts of goings-on."

He shrugged. "Dances aren't much fun without enough females. If the Purcell sisters had been there I might have found the place tolerable."

"Even England takes a bit of getting used to," Julia said.

"I thought Cologne was supposed to be nice," said Sarah.

"The place was all right. I don't like seeing people starve, that's all; even if they were the enemy a year ago."

For a year, ever since the sudden, unexpected Armistice, Sarah had waited for Gabriel to come home, imagining it would be like 1917 all over again. Something had gone very wrong. The comfortable companionship, the happiness of

1917 was no longer present. There was constraint, there were hesitations, unexpected halts in the conversation, as though they were strangers to each other searching for some common ground to discuss. When Gabriel said, "I'd better go home," no-one tried to persuade him to stay.

Frances pulled his shirt down from the creel and stood close to him as she buttoned it up. "You will change when you get in, won't you? Your trousers are still very damp."

He smiled wryly down at her. "Nanny Purcell."

She flushed. "I'll come over to the Rectory with an umbrella."

"There's no need."

"Don't be silly. Look at it outside."

He pulled on his jacket. "I'll see you all at dinner this evening. Oh, and by the way, would you mind very much not saying anything about the army, or Ireland? I don't suppose I shall have had time ..."

"Of course," they said.

From the kitchen they heard them talking in the hall, the sound of the front door closing. Then silence.

"Well," said Gwen, "you did wonder, didn't you?"

"There are times," Julia said, "when I could wring her neck."

I can't bear it, Sarah told herself. I thought he'd take me to the opera, show me Cambridge ... How can he go away again? To Ireland, of all places.

Bone china and tea in the drawing-room always brought back memories of childhood – those terrible Monday afternoons when Mrs. Mackenzie came over to have tea and go through the previous week's accounts. Even Sarah, too young to be involved in financial matters, could feel her sisters' terror as they sat with tea-cups perched precariously on their knees, faces growing whiter by the minute as they made polite conversation and waited for the account book to be brought out.

Nowadays Mrs. Mackenzie rarely came over to Hillcrest and formal tea was a thing of the past. But it was Mrs. Mackenzie who met Julia visiting Bill Roberts, the day after Gabriel had broken his news, and invited herself to tea with the Purcells.

"She told me she wanted to ask a favour," Julia said.

"I don't imagine I'm her most favourite person at the moment," Frances said. "I'll keep out of her way. I'll be in the stables if you want me."

"She particularly said she wanted to see all of us. It'll be about the sale of work for that Punjab mission of hers, I expect. You usually do the posters, Frances; you'd better stay."

So Frances stayed, sitting by the window, silent and withdrawn, contributing nothing to the conversation, while the others discussed the harvest festival, the unveiling of the war memorial, the relative merits of Hillcrest and Rectory cherry jam.

The scent of chrysanthemums drifted in through the French windows from the pots on the verandah. There was a smoky tang in the air, the smell of autumn.

"It's going to be a splendid year for colour," Mrs. Mackenzie said. "Look at that maple."

"I can never understand," said Gwen, "why one year should be better than another."

"I believe it's something to do with extremes in temperature," Mrs. Mackenzie said. "In Kashmir now the days are hot, with frosts at night, and the colours there are always magnificent. There's a special kind of light, too: hard to describe. You would like it, Frances."

She talked nervously and at length about the beauty of Dal Lake and the colours of the trees in the Shalimar Gardens, how delightful Kashmir was after the heat of the plains, while the Purcells sat patiently waiting for her to bring up the subject that had been her reason for coming.

Her hair had faded over the years so imperceptibly that

no-one noticed it was no longer fair but grey. Her face had gathered lines, her throat become crêpey, but she was elegant still and as beautifully groomed. Only when she was anxious, as now, was it possible to foresee her appearance in old age.

She rose to her feet at last. Her hands fluttered to her throat. "How foolish of me. I almost forgot . . ." Her hands played with the gold chains round her neck. "I suppose Gabriel has told you . . . about Ireland."

"Yes," Gwen said. "We were . . . well, surprised. We'd expected him to go back to Cambridge."

"So had we. When he went to Cambridge before coming down here we assumed . . . Mr. Mackenzie and I are sure that he is making a terrible mistake. He won't listen. We wondered – he does pay attention to you. Would you talk to him?"

"We have tried, Mrs. Mackenzie," Julia said. "He won't listen to us either."

Mrs. Mackenzie moved towards Frances. "Please, Frances. I know that in the past we haven't always agreed . . . but I believe we both want him to be happy. If you were to ask him to change his mind, he might . . . Frances, I'd give anything, I'd gladly give everything, to get him out of the army and keep him in this country."

Frances said slowly, "There's nothing I can do."

"You could . . . Frances, for Mr. Mackenzie's sake. He's growing old. He's tired, worn out by the war, and now Gabriel's making him wretched. Please."

Frances was looking at Mrs. Mackenzie with a strange expression, as if seeing her for the first time. "I have tried to persuade him, Mrs. Mackenzie. I don't want him to go to Ireland any more than you do. I've talked and talked." Mrs. Mackenzie's lips trembled. "All right. I'll try again, but I can't promise that it'll do any good."

"Well, well," Gwen said when Mrs. Mackenzie, trying to smile, had set off down the hill, "you've got her blessing at last. If you want it, of course, which you don't. I never thought I'd live to see the day."

"I think you're wicked, Frances," Julia said, gathering the tea-things onto a tray with unnecessary noise. "You've ruined his life. You know perfectly well you could stop him now if you wanted to. You ought to get right away and leave him alone. If you had a shred of decency you'd have done it years ago."

Frances flushed. "You've done nothing but criticise ever since you came home. Gabriel's always known how I felt. It was up to him to go on seeing me or not. It's not my fault." She stepped out onto the verandah. "I won't listen."

"It'd serve you right if you fell in love with a rotter," Julia shouted after her. "And I hope he treats you as you've treated Gabriel." She gave Sarah a shamefaced smile. "Don't look at me like that. It's time someone told her what they thought. Look at her. Independent, that's what she prides herself on being. One of these days she'll find out that she needs other people. She'll discover that other people matter too." She picked up the tray. "Open the door, will you?"

"I think Julia's being a bit unkind," Sarah said. "We matter to Frances. So does Gabriel, Julia knows that."

"Yes, of course," Gwen said, shaking the tablecloth out on the verandah. "Still, it's hard on Julia, isn't it, the way Frances is carrying on. When you think of Geoffrey and all that." She folded up the tablecloth and glanced over at Sarah. "I suppose you wouldn't have a word with him? He thinks a lot of you; you might have more influence than Frances at the moment."

"I could try. I don't imagine it would do any good." She watched Frances bending over the face of the sundial in the rose garden, fingering the letters: 'If you stand with your face in the sun . . .' "I'll see what I can do."

Chapter Twenty Four

Sarah had accepted that Gabriel would spend his first day home alone with Frances; had expected it. She had not expected him to be too busy to spend time with her. Now he was cutting short his leave; tomorrow he would be gone and she had not been on her own with him once. It was hard to bear and she blamed Frances for it.

She was lying by the maple in the long grass of the lower garden, squeezing out tears of self-pity and planning all the things she would say to Frances if only she were brave enough to say them, when a voice said, "You'll catch pneumonia lying on the grass like that," and she opened her eyes to see Gabriel standing between her and the sun.

"It's all right," she said. "It isn't wet."

"It can't be dry after all that rain the other day."

"Well, perhaps ..."

He sat down beside her, playing with a blade of grass. "I feel quite nervous. You're so grown-up now; not the Sarah I know. A young lady."

"I'm the same inside. I don't seem to have changed at all really."

"Disappointed?"

"No. I did think I'd feel different though."

He smiled at her. "In what way different?"

She hugged her knees. "I don't know. Just different."

"I don't think you've changed all that much on the outside either. What is it the Purcell family has against shoes?"

She looked down at her bare feet. "It feels so much nicer without them."

"That's the trouble with your family," he said impatiently. "It's always feelings that count."

She glanced warily at him. "Are you really going back tomorrow?"

"Not back. I'm going to London for the time being. It's easier to make the arrangements in London than from the back of beyond down here."

"When does the regiment go over to Ireland?"

"I'm not with the regiment any more. I'm being seconded. I'll go as soon as everything's fixed up, I suppose."

She said restlessly, "I wish I understood about Ireland."

"It's perfectly straightforward. They want to be independent; we won't let them. The Sinn Feiners won the last election; we outlawed the party. They wanted to take their case to the Peace Conference; we wouldn't allow it. Odd, that, when we're supposed to be sympathetic to small countries. I thought that was why we went to war in the first place."

"I thought they were traitors."

"You've been listening to my mother. How would you define a traitor?"

"Someone who betrays his country."

"And you think the Irish are betraying Ireland? By wanting to run their own country, in their own way?"

"Well ..."

"They're going about it the wrong way, though, which is what is causing the trouble."

"You might get killed."

"It's not likely. The Shinners are picking off the Irish constabulary and the Dublin detectives at the moment and ignoring the army. Besides, the work I shall be doing isn't dangerous – office work, mostly."

She was surprised. "Do you know what you're going to do?"

"Hugh Meredith's been trying to get me seconded to Ireland for weeks. He asked me to join his team some time ago. When I telephoned the other day he said he'd still be glad to have me . . ."

He stopped abruptly and reddened.

"The other day . . ." Sarah said. "So it was true. You did mean to go back to Cambridge."

"I thought about it. It was a mistake; one can't ever go back." He said gently, "Don't blame Frances, Sarah. She's never been anything but honest with me. It's my fault. I always knew that her work came first. I persuaded myself that she'd come round to my way of thinking in the end. I thought we could work something out somehow. It was stupid of me, I can see that now."

I hate her, Sarah thought. I'd like to pull her paintings out of the stables, one by one, and burn them all.

"Have you said good-bye?"

"Yes. She was busy with the latest masterpiece."

His voice was expressionless. Impossible to tell whether he was bitter or merely matter of fact.

"She promised your mother she'd try and persuade you not to go to Ireland."

"She did her best."

"It wasn't any good?"

"No."

"Suppose I asked you not to go?"

"Regretfully I should have to say no."

"If I refused to go to Oxford . . ."

"Only a silly little girl would make a threat like that, and as I know you're not a silly little girl . . ."

"I would, if it'd do any good."

"Well, it wouldn't."

She tried again. "What about politics? You can't carry on with politics in the army, can you?"

He was startled. "You know I haven't had anything to do with politics since before the war."

"Your mother used to think you might become prime minister."

He looked amused. "Oh, she did, did she? She obviously hadn't followed that thought through very thoroughly. What party did she think I would lead, I wonder? She used to think that Asquith was the devil in human form. No, I'm afraid that mothers are inclined to have an exaggerated idea of their sons' abilities. I was never likely to be prime minister, let me assure you."

"You've always been interested in that sort of thing, though. What about the Fabians?"

"That was when I thought I could put the world to rights. Before the war, when my illusions were still intact. When I thought that politicians, however misguided, were honest and well-meaning, that the British High Command was efficient and adaptable. The Somme finished all that. Do you know, until I saw their reports from the front I even believed what I read in the newspapers? I must have been incredibly naïve. So were most of my contemporaries, that's the extraordinary thing. Well, I have no illusions now; I know that whatever I do will make precious little difference to anybody. Perhaps that's one thing about going to Ireland – I might be of some use there."

"In what way?"

He thought for a moment. "I had an Irish friend once. He was on leave in Dublin in 1916 and got caught up in the uprising. He was a Unionist himself and came from a strong Unionist family so you wouldn't expect him to have any sympathy for the Shinners but by the end of the week he was on their side. There were murders – I'm not talking about the executions afterwards which, though badly mismanaged in my view, were different because there had been negotiations with the Germans – but cold-blooded murders of innocent people by the army. The officer in charge wasn't even reprimanded – he was promoted, in fact. It wasn't until a major who'd been there came over to England and saw Lloyd George that anything was done. Too late for the victims, of course. If I go over to Ireland now it seems to me that at least I have the rank to prevent that sort of thing happening. I disagree with their methods but I sympathise with their aims. Perhaps I can act as a friendly referee. Do you understand?"

"I suppose so."

"It's a sign of middle age, you know, moving from the general to the particular. When you're young you think you can improve the lot of everyone. The middle-aged know they're lucky to be able to help one or two. Old age must be when you realise it's only worthwhile concentrating on yourself. In which case, I'm old."

"Oh, Gabriel," she said miserably.

"I'm sorry. Let's talk of something else. I haven't congratulated you on your matriculation results. Were you surprised?"

"Amazed. I thought beforehand that I might go to pieces, not having done any exams before. I enjoyed them, as a matter of fact. They were fun."

He laughed. "Honestly, Sarah. Exams – fun!"

"Well – challenging then. I'm quite looking forward to the Oxford entrance. I go to Bishop Fox's for that, too."

"And you weren't homesick?"

"There was no need – I could have come home every evening if I'd wanted. It was just that your father thought Taunton was a bit far, with exams all day, and then Mrs. Steel offered to have me. Do you know Mrs. Steel?"

He shook his head.

"She belongs to the Victoria League. That's how she knows your mother."

"Which explains why we haven't met. Mother would keep us apart in case I came out with anything embarrassingly radical. Poor Mother. Sometime I must remember to ask her what she thinks of the Bolsheviks."

"I thought Mrs. Steel was nice, all the same."

"I'm sure she was. Just because she disagrees with my opinions doesn't mean she's not nice, does it?"

"She treated me as if I were grown-up. We had woman-to-woman talks. I liked that. People still think I'm about ten here."

"I'm sure that's not true."

"It is."

"Father looks on you as adult. I don't know what he will do without you."

"Oh, Gabriel. Think how much more time he'll have. I feel quite guilty sometimes at the hours I take up."

"He won't have anyone to talk to."

"What about your mother and Lucy?"

"That's only conversation. What about the things of the mind? You know how he loves a good discussion – argument. He only gets that from you."

"There's another year yet. Two perhaps – he says I'm still too young to leave home."

"I know. I don't agree." He picked up a leaf and frowned down at it. "Incredible colour, isn't it?"

"It's a maple. It's my tree; Mother and Father planted it when I was born."

He ignored her words and said, "I think it's important that you should get away from Hillcrest soon. I'd like you to go at once, but I suppose that's not possible."

She stared at him. "I don't see that there's any hurry."

His fingers tugged at the flesh of the maple leaf. "I don't know how to say this without hurting your feelings, but there are influences here that I think you should get away from as soon as possible."

"But Gabriel, what sort of influences?"

He said, after a long silence, "It's easy to accept the opinions of those around us without thinking very deeply about them. Sometimes it's hard to accept that people we love might be wrong. That's when it's important to go away and work things out for yourself, away from your own little world and ... the people round you. You're a clever girl, Sarah, but it's easy for clever people to isolate themselves from the ordinary world. I think it's particularly true of creative people, people like writers or ... or ... artists. They think they can manage without other people, that their own creations are enough. I don't think that's true. I don't believe that you can go through life ignoring other people without losing something in yourself – your humanity, for want of a better word. I would hate to see that happen to you."

The flesh of the maple leaf lay in fragments round him, twists of scarlet fibre resting on the grass.

"Do you understand what I'm trying to say?"

There was a band round her chest, making it difficult to breathe. She could not bear to look at his face.

"Yes."

He threw away what remained of the leaf. It landed among its fellows, a creamy skeleton like the backbone of a fish floating on a scarlet sea of fallen leaves.

"At last I've got that off my chest," he said, and tried to smile.

She wanted to cry. "Oh, Gabriel. Do you have to go away so soon?"

"I think it's best. I'm sorry – I'd been looking forward to seeing more of you. I'd thought we might go to the coast again. Do you remember that summer before the war when we took you to the sea?"

She looked back, years ago, to a remote golden past; felt his hair like silk against her cheek, the fresh evening breeze coming off the sea; saw the craggy Somerset coastline spread out like a map before her and the misty hills of Wales beyond the grey metallic sea.

"I remember."

"I've never forgotten. I used to go over it in my mind in the trenches, step by step, every foot of the way. That's what I thought I was fighting for in the end, and to give you a decent future, funny little thing that you were."

She thought of Achilles, striding above her along the brow of the hill, black against the sun. There was a lump in her throat.

"I went back yesterday," he said.

"All the way?"

"To the farm. It's not far; you can do it easily in a day. We were held back by the youngest member of the party, if you recall, who wasn't used to such expeditions. The farmer's wife gave me a cup of tea; she said she remembered us. Nothing had changed."

Was it lack of time that had stopped him from going beyond the farm, she wondered, or the memory of Frances on the beach?

"Do you remember the smell of bacon?" she said. "And mushrooms frying? I feel hungry at the very thought. They taste so much nicer out in the open. Like whortleberries."

"I'd forgotten about the whortleberries. If only you could have seen yourself – all over your face!"

"You ruined a handkerchief cleaning me up."

"So I did. Do you know, I don't think I ever got it back."

He lay back on the grass, hands under his head, smiling up at the sky. "Life was splendid in those days. We were all so confident."

She said hesitantly, "You were the one who said about pneumonia ..."

"I don't think I care whether I get pneumonia or not," he said, but he shifted round until he was lying away from her with his head in her lap. His eyes looked up into hers. "You think life is splendid, don't you, Sarah? At seventeen it is important ..."

"Yes, of course," she said, but she was not listening. She was taken aback by his sudden, unexpected closeness, surprised by the weight of his head pressing down on her thighs and terrified that he might hear the rapid beating of her heart. He lay with his eyes shut, saying nothing, and after a few minutes she realised by his even breathing that he was asleep.

His face was peaceful in sleep, the skin smoothed out. He looked much younger, almost a boy again. She had an almost irresistible desire to touch him: to stroke his hair, to run her fingers lightly over his face and feel the shape of the bones, the roughness of his skin. She wanted to hold him in her arms, to comfort him and give him hope.

Oh Gabriel, let me look after you. Please let me look after you. I could make you happy, I know I could. I'd do anything to make you happy. I don't care about Oxford or Hillcrest or the others. I don't even mind about your loving Frances best. I only care about you. You're all I've ever wanted.

She was trembling although she was not cold; her whole body shaking uncontrollably so that she was sure he must wake. He slept on unaware, his shirt breathing gently with the movement of his chest. A faint pulse beat in his neck.

Slowly the sun disappeared behind the elms; the warmth of the autumn afternoon cooled. Colour faded from the sky.

The ship's bell sounded from the verandah. Gabriel stirred, blinked, sat up.

"Oh, Sarah, I'm sorry. I'm rotten company at the moment, I'm afraid. I'm so tired."

"It's all right. It'll have done you good."

"What time is it? Lord, I shall have to go. You know what a stickler Mother is for punctuality."

"Yes."

"You'll still write, won't you?"

"If you want me to."

There was a trace of the familiar, crooked smile. "I shall have to hear about Oxford."

"May I still send you my writing sometimes?"

"Of course. It's coming along, you know. One of these days we'll be proud of you. I'm proud of you now."

He stood against the brilliantly coloured foliage of the

maple, irresolute, unable to drag himself away. In the distance the church struck the hour.

"Well," she said at last, "good-bye then."

"Good-bye."

He bent forward and kissed her, as he always kissed her when he went away: on the forehead, gently, without passion. Like a brother. "God bless you."

She watched him walk away up the slope of the garden, along the side of the dahlia bed, past the asparagus, until the protective hedge of yew hid him from sight.

Good-bye. God be with you. God bless you.

The only words that mattered, the words that could never be said. I love you. I love you. I love you.

1920

Chapter Twenty Five

A chill, damp wind whistled down the platform, searching out cracks and crevices, whipping round the extremities of those waiting for the train to depart. The Rector's face was sallow, pinched with cold.

"Don't wait," Sarah said. "You'll catch your death, as Annie'd say."

"You will take care, won't you?" Frances said. "Don't forget . . ."

All the way into Taunton she had gone on, about unaired beds, damp stockings, wet shoes, all the hazards that might befall. When it came to fussing, Frances could be worse than Annie.

"Oh, Frances! I'm not a baby."

Frances looked hurt. "If only you didn't have to change trains. I ought to come with you."

"I'll be all right," Sarah said quickly. Frances's expression made her feel guilty but the thought of arriving at Oxford with an older sister to hold her hand made her blood run cold.

Mr. Mackenzie put a hand on Frances's shoulder. "Of course you will. Frances, my dear, why don't I take you to Maynard's before we catch the train home? We can comfort each other over a cup of coffee."

Frances kissed Sarah through the open window. "Listen, mouse. If you decide it's not what you want – you know what I mean – it doesn't matter. We'll think of something else. So don't worry."

"Don't you worry either," Sarah said. "I've got Jane to keep me company."

Sarah was not the only one being fussed over. Jane's mother was as bad as Frances. Sarah, watching Mr. Mackenzie and Frances walk away along the platform, tried not to eavesdrop but could not help hearing.

"Have you got the letter? Do keep it safe. Remember, you must post it the moment you arrive at Aunt Mary's or your father will suspect. And, darling, please be careful when you come home – not a word about Oxford."

Curiosity got the better of good manners, although not until the train had drawn out of Taunton station.

"Doesn't your father know that you're going to Oxford?" Sarah said.

Jane shook her head. "He thinks I'm staying with my godmother. He doesn't approve of university for girls, you see. If my brother went to university why can't I? I said that to him, but he wouldn't answer. He thinks universities are only for boys."

"You'll have to tell him sometime."

"Yes, but Mother thinks he'll come round in the end – if he's presented with a *fait accompli*, you know."

"Suppose he doesn't?"

"I don't know what I'd do then. I think I'd die."

Something – not the words, which were too dramatic, but the way in which she said them – brought back the memory of Jess; Jess, five, six years ago, sitting one winter afternoon in the dim light of the potting shed while the rain drummed overhead, talking about her future as a nurserymaid. Everyone expected Jess to be grateful: a nurserymaid was infinitely better than going into domestic service. Only Sarah knew that Jess had hoped for a miracle to save her. For someone who had always wanted to teach, who had succeeded in getting little Davie Gould to read when Miss Ross herself had failed, being a nurserymaid was a dreary prospect. But Jess, like Gabriel, had discovered that miracles rarely happened.

Sarah stared out at the watery wastes of Sedgemoor passing behind her reflection in the train window. It was years since she had thought of Jess. Was she looking after babies still, Sarah wondered. It seemed quite wrong that she and Jess should now be so far apart when once they had been so close.

About all that Sarah and Jane had in common was the fact that they had taken their matriculation and Oxford entrance papers at the same school. It was surprising, Sarah thought as they changed trains at Reading, arrived together at Oxford, what confidence a companion gave. Perhaps if she had had someone with her when she went to Bristol, Frances, say, or Julia, she might have enjoyed school, settled better with the aunts. An odd thought that had not occurred before – her life might have been quite different.

It was the remembrance of Bristol that made her nervous, that had started the fear lurking at the back of her mind during the past few months, never expressed, barely acknowledged even to herself. Suppose Oxford were to turn out like Bristol – the object, so long desired, once achieved no longer desirable?

During the first day she held her breath waiting for disappointment. It never came. Instead, despite the strangeness of the buildings, the unfamiliarity of the other girls up for interviews with her and the nervousness brought on by the interminable waiting in the common-room before being called, she felt surprisingly comfortable and at home.

The girl in the next door bedroom obviously knew her way around. Enormously tall, she introduced herself to Sarah as Nell Holroyd, adding, quickly, "and please don't mention Dickens. Believe me, I've heard it all before." Nell's eldest sister had been at Oxford, her second sister was at the moment in the college; it had always been taken for granted that Nell, like Sarah, would come to Oxford. Nell was friendly, happy to answer any questions that Sarah cared to ask, to impart information.

It had never occurred to Sarah when contemplating university to wonder about her future fellow students. She had almost forgotten her desperate longing when younger to be like everyone else, having come over the years to accept the fact that she was different. It was true that in Huish Priory the oddity of the Purcell sisters was no longer remarked upon. If the Rector approved of girls going to university, the villagers said, then it must be a new-fangled idea that couldn't be altogether bad. Even Frances, once the source of endless gossip and speculation, had gained a kind of acceptance over the last year or two. Total strangers bought her paintings; art galleries round the country exhibited them. An article in an erudite magazine, circulated round the astonished villagers, described Frances as one of the outstanding painters of her generation. The inhabitants of Huish Priory gazed at her with the baffled awe of one whose ugly duckling has grown into a swan.

Lying in bed that first night, icy feet tucked up into her nightdress, hot water bottle clutched to her stomach, Sarah thought how different Oxford was from Hillcrest. No mention of paint in the conversation, of Picasso or Matisse. How many of the other candidates had heard of Kandinsky or knew anything about Klee? In these surroundings it would be her sisters who would be out of place. The knowledge struck her like a revelation.

Examinations did not frighten Sarah. She found them challenging and forgot, when writing, about the absent examiner. But interviews were a different matter. Impossible to banish all thought of the interviewer from mind when she was sitting squarely in front of you.

Listening to Antony's lessons in the Rectory study, Sarah had come to realise early in her schooldays that there was more to learning than the passive absorption of knowledge.

The Rector had always expected her to think for herself; he had encouraged her to form her own opinions, support and defend them, made sure that she could argue logically and calmly.

To have to discuss the essays she had written in the entrance papers and explore her ideas in greater depth with the tutors was no more than she would have expected to do at the Rectory. Her tenseness disappeared. She began, to her surprise, almost to enjoy herself.

Until the last interview. At first it seemed no different from the others; the questions not as searching perhaps but no more than that. Then the tutor picked up a book from one of the piles on her desk, opened it and handed it to Sarah.

"Read that."

It was a collection of poetry. Sarah was so taken aback, so unprepared, that she reached the last line of the piece marked without taking in a word, without realising what it was about.

After the first interview she had forgotten to be nervous. Now she could feel fear coming back, a hollowness in the pit of her stomach, like hunger. She had not expected this to happen, had assumed questions would be on her entrance papers or on general subjects, not on something specific.

What was she supposed to say? What did the tutor want? A résumé? Analysis? Criticism? If she could recognise the poem it would help, if she could guess the poet.

"The one red leaf, the last of its clan,"

Was there something at the back of her mind, a faint memory?

"Hanging so light, and hanging so high,
On the topmost twig that looks up at the sky."

The tutor's fair hair was scraped back over her ears and twisted into a tight bun that made the ice-blue eyes seem more prominent than they were. They fixed on Sarah like fish's eyes, cold and wet.

"Well?"

"It's about loneliness," Sarah said and stopped. Was it? Or was it nothing more than a descriptive piece? Surely not. They wouldn't do that, would they – try to trick her? There must be some inner meaning that she would be expected to find.

She stumbled on, her brain working on two levels, outwardly talking, trying to find meaning in the lines, while inwardly she contemplated the failure of her hopes and those of Mr. Mackenzie.

When she returned, still shaking, to the common-room the others waiting there were reassuring.

"I don't suppose it was as bad as you think."

"It's only one interview after all. How did you do in the entrance papers, do you know?"

Sarah blew her nose. "All right." It was an understatement: Mr. Mackenzie had been told that on her entrance papers alone she could expect to gain a place. That knowledge was not much comfort now, for what effect would a failed interview have?

"Well then," Nell said. "My sister always liked her when she was up, you know. And she's very good at tennis."

"She probably hypnotises her opponents into playing badly," Sarah said sourly, picturing those icy eyes on the other side of a tennis-net.

She went into her interview with the Principal with a fatalistic sense of calm.

"So you come from Somerset," the Principal said. "A lovely county I always think. Do you know Alfoxton, I wonder?"

"Wordsworth's home, do you mean?"

"Yes, indeed."

The Principal had stayed at Holford the previous summer while making a study of the Lake poets. She had spent several weeks exploring the countryside around...

It was during discussion on the Quantocks scenery and its effect on the poets that Sarah suddenly remembered. Hadn't

Dorothy described a solitary leaf hanging from the topmost branch? In which case, remembering how much inspiration they had both derived from her, it must have been Wordsworth or Coleridge who had written...

Surprisingly her failure no longer mattered. She had done her best: it was all she could do. It was more interesting now to discuss Wordsworth, to wonder whether, had he been less moody, less given to solitary nocturnal rambles, he would have been accepted by the locals and allowed to remain at Alfoxton. Sarah, thinking of the villagers she knew and remembering spy scares in more recent times, thought not and was secretly relieved. How terrible to have had a literary shrine made of the Quantocks, to have strangers gaping and crowding along the empty paths she knew so well.

The Principal had followed in the footsteps of the Wordsworths and Coleridges, as the Purcells and Mackenzies had planned to do – would have done, had it not been for the war – from Alfoxton to Watchet and Minehead and on to Porlock. How different might Kubla Khan have been with different scenery – if Coleridge had already removed to the Lake District for instance? Once one began to wonder about the influence of scenery there were endless trails to follow ... Thomas Hardy ... the Brontës...

Sitting in front of the fire in the untidy book-filled room, Sarah lost all sense of time and place. She might have been back in Mr. Mackenzie's study, discussing writers and writing with Gabriel. She could happily have sat talking for ever.

Chapter Twenty Six

It was the thought of Jess that made Sarah think of money. She was shocked. Why had it not occurred to her before? If the Purcells had been unable to afford a week at Budleigh Salterton how could they possibly manage to send her to Oxford for three years?

"Not so chirpy these days, are we?" Annie said. "You'd tell old Annie if there was anything wrong, wouldn't you, love?"

"I'm all right," Sarah said. It was not something she could discuss with Annie; nor the Rector either, when he had devoted so many years to her study.

Frances was sitting by the table at the far end of the verandah. Elbows resting on the bleached wood, head supported by her hands, she gazed abstractedly out over the lawn. She looked pinched and cold, but it was no use telling Frances that she should be wearing a coat. She turned her head and smiled faintly at Sarah.

"Come and sit down. Still dreaming of the distant spires?"

Sarah pushed away the trays of plants sheltering behind the glass and pulled up a chair.

"I suppose so. It's very odd, though. I always knew about the spires – people talk about them and Oxford in the same breath after all – but they were a terrible disappointment when I actually saw them. I don't know why. I think they were so sharp and cold, not a bit like church towers round here which are pink and square and friendly somehow."

"Expectation's a funny thing. I didn't expect London streets to be paved with gold but it had never occurred to me that they might be so horribly dirty."

Sarah pulled a pot of sweet peas towards her. Tall and straggly, the plants flopped between the supporting sticks, tendrils clutching the canes and each other in a confused mass of greenery. She began teasing them apart, disentangling stalks and leaves, while she wondered how to bring up the question of expense.

Frances, after the brief moment of animation, had returned to her contemplation of the garden. In repose her face was sad. It was her vitality that brought her to life and gave her beauty. There was no life there today. What was she thinking, Sarah wondered. There was no way of knowing. Frances sat in the wicker chair, remote, withdrawn, as if she had forgotten Sarah's presence. In the glass her pale face was reflected against the dark background of the hedge, blurred and uncommunicative.

"I've been thinking about Oxford," Sarah said at last, her fingers playing with the sweet peas, "and thinking ... If I do get a place ... isn't it going to be awfully expensive?"

"Oh yes. Quite awful, I expect. Don't be silly. We'll manage."

"Yes, but ... if we haven't got the money."

"Dear Sarah, if I was able to go to university for three years then you most certainly shall."

Sarah stared. "You went to the Slade."

"Which is part of London University. That surprises you, doesn't it? You didn't think I was bright enough. Well, I'm not, of course, not in the way you are, but I got the training I wanted and needed. That's what counts. If I'd had my way both Julia and Gwen would have gone, too. I don't know that the Slade would have helped Gwen all that much, but I still think it's a great pity that Julia took fright — although I suppose she'd have left for France in the middle anyway. I've told you before: education's what counts. You've got to be able to be independent, and a decent education's the only way to achieve it. Unless you have money, I suppose, which we haven't. And I think it's much more important for women than for men, though not many people would agree with me. Men can always go off to Canada or Australia somewhere like that. That's where I get so cross with the Mackenzies. It was always taken for granted that the boys would go to university. No-one gave a thought to Lucy. All right, I know she's not brilliant but what sort of education did she get? I can't understand Mr. Mackenzie sometimes, really I can't. Music, sewing, the classics — just skimming the surface. What's she fit for?"

"She does a lot in the parish," Sarah said. "She's taking more and more of the parish work over from Mrs. Mackenzie these days."

"Oh yes. She's very useful; that's the trouble. She's not the Rector's wife though, is she, only his daughter. There'll come a time ... and then what? Oh, look, we're talking about you, not Lucy. What I'm trying to say is that Oxford's the greatest opportunity you're likely to have in your life and we'll not let you miss it."

"But Frances, the money ..."

"We're not poor, you know. In fact, if we'd done what everyone wanted when Mother died and lived in a rented house in Taunton we'd be comfortably off. Living here, in a place this size, we have to be careful, that's all. We don't live beyond our means, we manage to save most years, but you never know when a gale might not take off the roof or one of us fall ill. As for your going to university, we've been putting money by for that ever since Mr. Mackenzie mentioned it as a possibility."

Sarah stared. "No-one ever told me."

"No. I suppose we should have. You've always been so

much younger than the rest of us we forget you're growing up. I can't really believe that you're older than I was when Mother died."

She shivered and hugged her arms close to her. "It's not just the education, either, although that's important. When we were growing up life was such fun – here, London, visiting Gabriel in Cambridge. Nowadays everything's so ... so serious. Nothing's been the same since the war. We're a grey lot ourselves. You ought to get away and be with people of your own age. I'd like to think you could be as happy as we used to be."

"I'm happy now."

"Besides we're relying on you to bring fame to the family name."

"What do you mean? What about you?"

"I'm no good," Frances said. She sounded tired. "I'm finished. I haven't painted anything decent for months. I tell myself that I'm going through a bad patch but I don't believe it. They do happen sometimes, not bad patches exactly but sort of fallow periods. Do you remember 1917? That wonderful summer? I painted and painted and painted and everything came out better than I dared hope. After Gabriel went back I didn't paint at all for weeks but it didn't seem to matter. I felt – it's difficult to explain – out of this world somehow. Now I just feel exhausted. And frightened. Suppose I've reached my peak? I've never thought of such a thing before. Now I think about it all the time. I keep remembering Mother. I wouldn't want to paint at all if all I was doing was mediocre stuff, but there doesn't seem to be any point in living if I can't paint. I might as well be dead."

"Oh, Frances," Sarah protested. She had never known Frances like this; she was alarmed. "It'll be all right. It's just a matter of time. Look at Julia. She's drawing again. We never thought she would."

"The divine spark, Gabriel used to call it," Frances said; she was talking to herself, oblivious of Sarah. "Nothing to do with you. Either you've got it or you haven't but if you have ... I always thought that if you disciplined yourself and worked hard enough it would go on for ever. Perhaps it doesn't. Perhaps it just peters out. What do I do? How can I go on?"

"It sounds a bit like writing," Sarah said cautiously. "The spark's like an idea. You can't do anything without that. And yet, somehow, it's there or it isn't."

Frances brought her gaze back from the garden. "Where do your ideas come from?"

"I don't know. They're suddenly ... there. Some kinds anyway." She wanted to distract Frances from thoughts of painting but found it difficult to explain something that she barely understood herself.

There were the day-to-day stories, the entertainments which could be summoned up at will. There was no art there: it was what everyone did. Or did they? She had wondered for years: if other people did not tell themselves stories, what did they think about? How else did they fill the long spaces of time?

Then there were what she herself termed 'proper stories', not divertissements but stories to be plotted, developed, drafted and redrafted, worked on while she resisted the temptation to slip back into the fantasy tales in her head that required so little effort.

She said thoughtfully, "I don't know where ideas come from. They just arrive. They're only ideas though. I can't always turn them into something worthwhile, and when I do it's jolly hard work. Agony sometimes."

"Of course. It's worth it, though, when you do succeed. Isn't it?"

"Oh, yes," Sarah said. For a moment she glimpsed, without being able to put into words, the similarity between Frances and herself, and the idea of people as instruments, through paint or words, of creation. She was suddenly afraid. "Do you mean ideas might stop coming?"

"I don't know. I never thought I wouldn't be able to paint."
Frances dragged herself away from the unhappy contemplation of her own thoughts and regarded Sarah curiously.
"Aren't you odd? Why have you never mentioned your writing? Do you realise that if it weren't for Gabriel we wouldn't have known?"

"I don't mind your knowing. It's not something I can talk about, that's all. Don't worry, I'll tell you soon enough when I get something published. I'm not very good yet, you see."

"Gabriel has always said you showed a lot of promise."

"Has he?" She was surprised, and pleased. "He's always terribly critical. All the things I send him come back covered with comments. He's usually right, that's the trouble. Still, when he says something's coming along ... and he's always encouraging. Now that I've got the Oxford entrance over I should have more time. Do you know, they have newspapers and magazines one can try and write for at Oxford? Think what good practice that'll be. If I go, of course."

"I keep telling you, you're going. We'll manage. There are all sorts of people we can go to for help if we need it. Sir James, for instance. He offered some time ago. The Mackenzies, too, though they've done so much for us already – and it's Mrs. Mackenzie who has the money in that family – that I'd rather not take their help unless we have to."

"I can't believe Sir James . . ."

"There's not much goes on in the village that he doesn't have a finger in. Then there are the aunts. Julia and I were talking about them the other day. They were going to pay for your board and uniform in Bristol, you know. They wanted to pay the school fees too, but I wasn't having that. I think they've got quite a good income. It's the capital they can't touch; it goes to a home for fallen women or something when they die. I'm sure they'd be delighted to help out if we asked them. You made a very good impression when you stayed there." She smiled at Sarah. "So you see, there's no need to worry. I'd prefer to manage on our own if we can, that's all. If the worst came to the worst I could always teach. I wouldn't mind working in Taunton a couple of days a week if I had to. It'd be very good discipline, I expect."

"Oh, Frances!" She was not sure whether to laugh or cry at the picture conjured up, at the thought of the school population of Taunton put off painting forever by Frances's well-meant, crushing criticism.

"There is one thing, though," Frances said. "There'll be people there with a lot more money than us. Even the Slade had society people, girls especially. There are sure to be more at Oxford — girls who can spend money on fashionable clothes and all sorts of luxuries while you have to save for books and things like that. You'll have to be prepared . . ."

"I know," Sarah said. "I shan't mind." She blushed at Frances's glance. "Well, if I do it won't matter."

"I tell you what we could do," Frances said. "Gabriel bought a picture of mine last time he was home. Don't look so surprised. I didn't want him to pay for it. He knows perfectly well I'd give him any painting he wanted. He said this was for a wedding present and insisted on paying. He was very peculiar about the whole thing and offended and — oh, I don't know — rude. We had a terrible row in fact. In the end I took the money to keep him quiet. I don't want it though and I certainly shan't spend it. Why don't you have it? You can use it for extras when you start in Oxford — clothes, books, whatever you want. I'd like that and I know Gabriel wouldn't mind. You can tell him if you like. How about it?"

"Well, if you're sure. Thank you very much." She was curious. "Don't you mind having rows, Frances? I'm sure I'd hate it."

Frances was playing with the sweet peas. She sighed. "I don't know. In the old days it was different. Sometimes — it probably sounds ridiculous to you — it was good fun. Having fights, I mean. Exhilarating, in a funny sort of way. It's not like that any more. The trouble with rows is that they get out of control. You start saying all sorts of things you don't mean to say, bringing up things that should have been forgotten years ago. Silly things. I hear myself doing it and somehow I can't stop. And then Gabriel . . ." She stopped. Sweet pea tendrils curled round her fingers like thin green rings. She asked casually, not looking at Sarah, looking down at the peas, "Have you heard from Gabriel lately?"

"Oh yes. He was very relieved I liked Oxford. He was afraid I might decide against it, like Bristol."

Frances smiled faintly. "I think we all were."

Sarah thought of Gabriel, trying to help the Irish in Dublin, disillusioned, lonely, sad. "I don't think he's very happy in Ireland, is he?"

Frances's smile faded. After a moment she said, "I don't know," and stood up. "How I hate the winter. I'd give anything for some sun."

"Never mind. It'll be Easter soon. We always manage to eat out over Easter. That's when the summer starts, don't you think, with meals in the garden?"

"Perhaps. It's much too cold now though." She sighed. "I'll see if there's any forsythia in flower. It might cheer us up a bit."

Do we need cheering up, Sarah wondered in surprise, watching Frances walk across the lawn. Frances's skirt barely touched her waist and hung loosely on her hips. She must have lost a great deal of weight. But Sarah was back in Oxford, thinking of that awful interview with the tutor, the talk with the Principal, of Nell. Dreaming . . .

Chapter Twenty Seven

Sarah had not expected to receive the results for several weeks. When only a day or two after the conversation with Frances, Lily came into the living-room holding a telegram as if it might burst into flames in her hand, she was astonished.

"For me? But I don't know anyone . . ." She stared down at the paper. "It's from Oxford. Fancy sending a telegram." The black marks on the paper turned into letters, formed themselves into words. "I've got an exhibition."

"You are a funny girl," Frances said, "standing there like that. Aren't you pleased?"

"I suppose so. Yes, of course. I can't believe it, that's all."

"It'll be horrid without you," Gwen said. "You at Oxford, Frances in Spain – Julia and I'll rattle around here."

Sarah stared. "You're not going to Spain, are you, Frances? Whatever for?"

Frances shrugged. "I don't know. I haven't painted anything worthwhile for months. You know that. I thought perhaps the sun – the light's better there. And I've got friends in Spain at the moment. It was Julia's idea. She thinks I should get away for a bit. Only . . . I don't know. I can't decide."

What could she say? 'It's so far from Ireland.' No, she mustn't say that. And Frances so uncertain, so indecisive. The fact that Frances was unable to make up her mind was almost as surprising as the thought of her going away from Hillcrest.

"I don't suppose I shall be able to stay away for long," she said at last. "I'll probably be like Frances at the Slade – dashing home whenever I can."

Annie took the news calmly. She was making bread, pummelling the dough on the table with large, capable hands, and did not even look up when Sarah burst into the kitchen.

"If that's what you want, love, I'm glad. You've worked hard for it, I'll grant you. Don't let it go to your head, that's all I ask. I wouldn't want you to get too clever to talk to us country bumpkins."

"Oh, *Annie*! As if I would."

Annie banged the dough down on the table. "Stranger things have happened. Foreign travel takes people funny ways."

"But Annie, Oxford isn't *abroad*. Not like Spain – did you know Frances was thinking of going to Spain?"

"Well, now," Annie said cautiously. "She has mentioned it. Nothing definite though, is there? If you ask me, Miss Frances isn't herself these days. 'Doesn't know what to do'. 'Can't make up her mind' – Miss Frances, of all people." She gave the dough a final thump before taking it over to the dresser to rise. "At least we know what you're doing, don't we, love? The Rector was pleased, I'll be bound."

"I haven't told him yet," Sarah said. "I'll go now. He should be back from school."

The hall was empty, the Rectory deserted. Her footsteps echoed on the polished boards.

"Hello," she called tentatively and waited.

The sun came through the leaded lights of the window on the stair landing. Specks of dust danced in its beam. The coloured panes turned the light into amber and red patterns on the floor. She let her gaze wander round, as if seeing the hall for the first time, or the last: the richly patterned Indian rugs, the oak panelling, the Mogul prints on the walls. Through the open doors she could see into the rooms: the desk in the morning-room neatly stacked with Mrs. Mackenzie's lists and papers; furniture grouped together in the drawing-room, waiting to be turned out; the study . . .

She caught her breath. Mr. Mackenzie stood in the doorway watching her.

"You made me jump," she said. "I thought you were all out."

"Mrs. Mackenzie has . . . a migraine," he said slowly. "She has gone back to bed. What are you doing here?"

She held out the telegram. "It's come."

He looked down at the envelope without moving. "What is it?"

"From Oxford."

He glanced at her face. "My dear child. Why didn't you say? Come in."

In the study he swept the papers on his desk into a drawer and pulled up two chairs. "Let me see."

He sat hunched up in the wing chair, staring down at the paper held between thin, bony fingers. The pale north light from the window fell sideways onto his face, accentuating the deeply cut lines between nose and mouth, the hollows under his cheekbones and dark shadows round his eyes.

Why, Sarah thought and was shocked, he's an old man.

She waited for him to speak, but he said nothing.

"I'm sorry I didn't get a scholarship," she said at last.

He smiled at her then. At least his smile was not old. It reminded her of Gabriel. Would Gabriel look like his father now in years to come?

"My dear child, I'm delighted that you gained an exhibition. I'd have been pleased with a mere entrance, you know – you're very young, after all."

"It's all your doing. How can I thank you?"

He shook his head. "It is I who should thank you. You have given me great happiness over the years, my child; more than you could realise. And now, at last, you are leaving us ..."

The sadness in his voice made her throat ache. "There'll be the holidays. I can still come here, can't I? Please."

"We always hoped that you would look on this house as a second home. You know you are welcome at any time. But you must go forward, my dear, and leave us behind. I'm afraid that here in the Rectory we live too much in the past."

He gazed into the distance with unseeing eyes. "It's a strange thing, memory. I can see you coming through that door on your first morning as if it were yesterday. So eager for your lessons; such big eyes taking everything in; so small that we had to raise you up on cushions to let you see over the table."

"I remember the cushions. They were dreadfully scratchy. I couldn't sit still. You kept on telling me not to wriggle."

"So different from Gwen. She was only interested in her plants and her drawing, even then, while you – you soaked up knowledge. Like a sponge, Gabriel said, that summer he gave you lessons. I've been sitting here thinking – such happy times."

"Gabriel will be pleased about Oxford, won't he?"

"He's always been fond of you, ever since that afternoon you fell asleep on his lap. Ten years ago. If I had known then ... I would have refused your mother, you know, if it had been possible. Guardian of four young girls, unknown young girls – how could I take on such a charge? There was no time. Your mother was dying. However inadequate I felt, however unworthy, there was no-one else. Now when I look back I can see how fortunate I have been. Truly blessed. You remember the old hymn, 'God moves in a mysterious way'? I tell myself that. One should never doubt ... never ask why ..."

His voice died away. She could think of nothing to say. He turned towards her with a bleak smile. "My dear, you must excuse me. I'm a little tired this morning – and your sisters will be wondering where you are."

She stood up. "I'll go now. I just wanted you to know." For a moment, looking into his eyes, she longed to hug him. Unaccountable shyness held her back.

She walked home through the churchyard. The day was fresh still, the frost of the night before cold on the breath. On the grass and bushes the white rime was turning into drops of water that sparkled in the sun.

Only now did she realise the full measure of her success. For years she had worked towards Oxford, since the time when the word was nothing more than a name to her: because Mr. Mackenzie had held it out as a goal, because she enjoyed a challenge – more than anything else perhaps because it would be an achievement to set beside her sisters' talents. Now she saw that entrance to the university offered her all that she had ever looked or longed for in the years of her growing up – intellectual companionship, independence, but more than that it offered her opportunities and a way of life that she had never dreamed existed.

She turned to look back before she went into the house. Beyond the church in the far distance the gently rounded, russet-coloured Quantocks rose like dream hills out of a white mist.

TINKER'S CAREER

Alison Leonard

Tina was just a baby when her mother died. Now fifteen, she's determined to find out more about her. So, finding a photograph of her parents' wedding she sets out in search of her mother's family and the truth. But the truth – and with it the meaning of Tinker's Career – turns out to be even more devastating than she'd feared...

"Told at a heady pace with wonderful real, absorbing characters."
The Guardian

"Strong stuff."
TES

THROUGH THE DOLLS' HOUSE DOOR

Jane Gardam

Claire and Mary love the dolls' house and its curious assortment of residents: the outsize Dutch doll, Miss Bossy; the General and his troop of Trojan soldiers; the miserable Small Cry; the mysterious Sigger . . . But little do the girls know of the extraordinary lives and adventures, past and present, of this resourceful band and the marvellous stories they have to tell.

"An original story . . . wry and funny, and full of a sharply poignant sense of the passage of time."
Jill Paton Walsh, Books For Keeps

FRANCIE AND THE BOYS

Meredith Daneman

When "quiet and vague" Francie unexpectedly gets a part in the Sir Henry Dubbs' School for Boys' play, she finds herself thrust into the limelight – and into ever closer contact with those "weird, wild alien beings commonly known as boys". Over the following months she learns a great deal about acting – its illusions and heartbreaks, as well as its glamours – and discovers that in drama, as in life, things are not always what they seem...

"Delightful heroine... The emotional understanding of the girl is most striking."
Evening Standard

"Charmingly believable... Affectionately comic."
The Times Educational Supplement